The Old Measure

The Old Measure

An Inquiry into the Origins of the
U.S. Customary System of
Weights and Measures

Jon Bosak

Pinax Publishing
Ithaca, New York

*The Old Measure: An Inquiry into the Origins
of the U.S. Customary System of Weights and Measures*
ISBN 978-0-615-37626-4

First Edition
August 2010

Copyright © 2010 Jon Bosak
All rights reserved

Pinax Publishing
1448 Trumansburg Road
Ithaca, New York 14850

Designed and illustrated by the author
Printed by Lightning Source
Distributed by Ingram Publisher Services

Dedicated to the memory of

U.S. Secretary of State Thomas Jefferson, who labored four years to design a decimal replacement for the U.S. Customary System of weights and measures

and

U.S. Secretary of State John Quincy Adams, who spent four more years explaining the system we decided to keep

Pleas' hyt yor good Lordshypps ... to understonde that afore the Conqueste sondrye kynges aprovyd and provyded how & by what maner of weyes they myghte beste & longeste endure; for a verey trowthe, how for to make a lib. [pound] Troye; to make that by Troy weyghte & measure, and owte of that make that by trewe weyghte and measure; and owte of that make the habar de payse pownde as ye shall hereafter knowe. In no wayes they cowde not fynde hyt but by means of whete cornes. Then contryvyd they how many whete cornes shulde goo to the Troye once [ounce]; and after concludyd that xii Troy onces shulde make the lib. Troye. Soo that ye may undyrstond that the once dyd not make the whete, but the whete made the once ...

And xii of these onces made the trewe pynte at thys tyme, and at thys tyme mayntayned the seyd trew pynte, that ys the kynges trew standard in hys Eschequer, that makythe all maner of measure, viii pyntes to the galon; that longyth to whete and lycour from the smalleste assyse unto the gretyste; as well as for all maner hother grayne that shuld be mesuryd as well as whete. And as that once made the pynte for grayne, it seruythe as well for wyne; that ys as myche to say the Gascoyne tonne, from the leste measure to the moste; and as well for all maner of hother lycowrs that conteyne measures in thys kyndde, they to be alowyd theyr nombyr, as the whete afore made the Troy lib. And all maner of measurys owte of that for a generall rule & for trowthe ...

Hereby may ye perseyve & perfytelye understond that there ys but on' weyghte & on' measure, accordyng to Magna Carta, that ony man owghte to serche or to bye or sel bye. So that ye may perseyve that Whete Cornes rulythe all maner of weyghte & measure. And so dothe none hothyr thynge in all the Worlde.

Exposure of the Abuses of Weights and Measures for the information of the Council
by John Colyn, a merchant of London (1517)

Contents

Preface vii

Acknowledgements xi

Introduction: Postdecimal numbers 1

1 The Northern foot 3
   Customary land measure 3
   Customary area measure 4
   Modern equivalents 5
   The Northern foot 6
   Archaeology of the Northern foot 10
   Survey foot vs. international foot 11

2 The troy pound 15
   The destruction of the troy standard in Britain ... 16
   The survival of troy weight in the United States ... 17
   The troy system 18
   Modern uses of troy weight 19
   The ancient history of troy weight 20
   Egyptian origins of the troy standard 21
   Troy weight in the Iron Age 23
   Spread of the troy standard 25
   Troy weight in Saxon England 25
   The haverdepois pound 26
   The pound of Cologne 29
   Troy weight in the Arabic world 30
   Troy weight in India 35
   Troy weight in the Far East 36

3 The wine gallon 41
   The customary gallon system 41
   A bushel and a peck 43
   Cask measures 44
   Oldest statute definition of the wine gallon (1496) ... 45
   Queen Anne defines the wine gallon (1706) 46
   History of the wine gallon in England 46
   European gallon cognates 48
   The Carvoran measure 50
   The native gallon 52
   Before the Romans 53
   Back to the land: the 10-gallon anker 54
   The Ur-barrel of 32 gallons 55
   Babylonian pints and quarts 57
   The tomb of Hesy 59
   The wine pint in Greece 62
   The Hebrew gallon 62

4 The Winchester bushel 67
   The Winchester system 68
   Winchester barrels 68
   Winchester measure for liquids 69
   Modern definition of the Winchester bushel 70
   Original definition of the Winchester bushel ... 71
   The tower pound 72
   Ancient lineage of tower weight 74
   Weights and densities 77
   Winchester measure as a Saxon import 78
   The great confusion 79

5 The avoirdupois pound 87
   The avoirdupois system of weight 88
   The word *avoirdupois* 89
   The downfall of the bushel weight systems 91
   The good news: pounds and gallons 92
   The decimal pound 93

6 King Henry's arm 95
   Theory 1: roundoff error 96
   Theory 2: densities and cubic inches 96
   Theory 3: inches and cylinders 98
   Theory 4: ounces and cubic feet 100
   Assessing the theories 102

7 Conclusion — 105
  Metrication: a mixed bag 105
  Living with the status quo 106
  Keeping in sight the historical objective . . 108
  A few diffident recommendations 109

Appendix A. Grain weight as a natural standard — 111
  Generic wheat and barley 112
  Behavior of specific varieties 114
  The Cornell data 116
  Which wheat? 119
  Export wheat 119
  The IWWPN 119
  Conclusions from the statute definitions . 122
  The imaginary wine bushels of Henry VII . 122
  "A pint's a pound" 125

Appendix B. The English statute basis of U.S. weights and measures — 129
  The English statutes in U.S. common law . 129
  English statutes relevant to standards of weight and measure 130

Glossary of unit names — 139
  Abbreviations 139

Frequently referenced sources — 151

Works cited in the text — 153

Index — 161

Preface

On a road trip in 1983 I saw a sign giving a distance in miles and wondered: why is a mile 5280 feet long? How did it get to be just that long? And why didn't we divide the mile into 5000 feet, as I knew the Romans did, or by some other sensible number? In short, how did the system of weights and measures we use in the United States, officially known as the U.S. Customary System, come to be the way it is?

I knew about the metric system, of course; a year apiece of physics and chemistry in high school and then another year apiece in college had left me with a working knowledge of what's known formally as the SI, and in fact I knew a fair amount of its history. But all I knew about the history of the system I use for maps and groceries was that the yard is the length of King Henry's arm, and I wasn't clear on which Henry.

I soon discovered that establishing the history of how the Customary System came to be wasn't going to be easy. For some reason, none of the archaeologists seemed to be much interested. And the few popular books on the subject of weights and measures were either tiresome screeds urging adoption of the metric system or shallow paeans to "natural" units such as the palm, the hand, the span, and the cubit (which is absurd—all measures are unnatural).

On further investigation, it became clear that the dearth of substantial historical material on the subject of weights and measures was largely due to the rise in the latter half of the 19th century of something called Pyramidology. The 1864 publication of Charles Piazzi Smyth's *Our Inheritance in the Great Pyramid* spawned a movement that lasted well into the 1930s and helped inspire the early years of the Jehovah's Witnesses. A suitably scathing brief account of Pyramidology and its connection to millennialism can be found in Martin Gardner's *Fads and Fallacies in the Name of Science*.

Pyramidologists believe that the Great Pyramid of Egypt, if properly measured by something called the Pyramid Inch, reveals in its features the history of the world, both past and future. Figure 1 shows a typical example.[1] The Pyramid Inch was held to be the ancient ancestor of the inch we use today.

This might have been harmless enough, but beginning around 1880, Pyramidology was seized upon by opponents of the metric system who believed that the Pyramid revealed our Customary units of measure to be divinely inspired. Gardner quotes the words of a song from *The International Standard,* a publication of this movement, the fourth verse of which runs:

> Then down with every "metric" scheme
>   Taught by the foreign school,
> We'll worship still our Father's God!
>   And keep our Father's "rule"!
> A perfect inch, a perfect pint,
>   The Anglo's honest pound,
> Shall hold their place upon the earth,
>   Till time's last trump shall sound!

A particularly bizarre 20th century version of this doctrine called the Anglo-Israel movement held that Anglo-Saxon and Celtic peoples are descendants of the ten lost tribes of Israel—and a heavily manipulated history of our weights and measures played a central role in this.[2]

Given the bad odor surrounding the subject of historical metrology (as it's properly called), the reluctance of modern historians to trace our standards farther back than the medieval era is hardly surprising. And the innocent but unfortunate tendency of early 20th century archaeologists to propose theory ahead of the data soured a couple of generations of later archaeologists on the subject as well.

Two otherwise respectable 20th century amateur attempts at historical metrology suffer from defects that render them less than reliable. E. Nicholson's *Men and Measures* (1912) covers a lot of ground and is certainly correct in tracing many premetric units

Preface

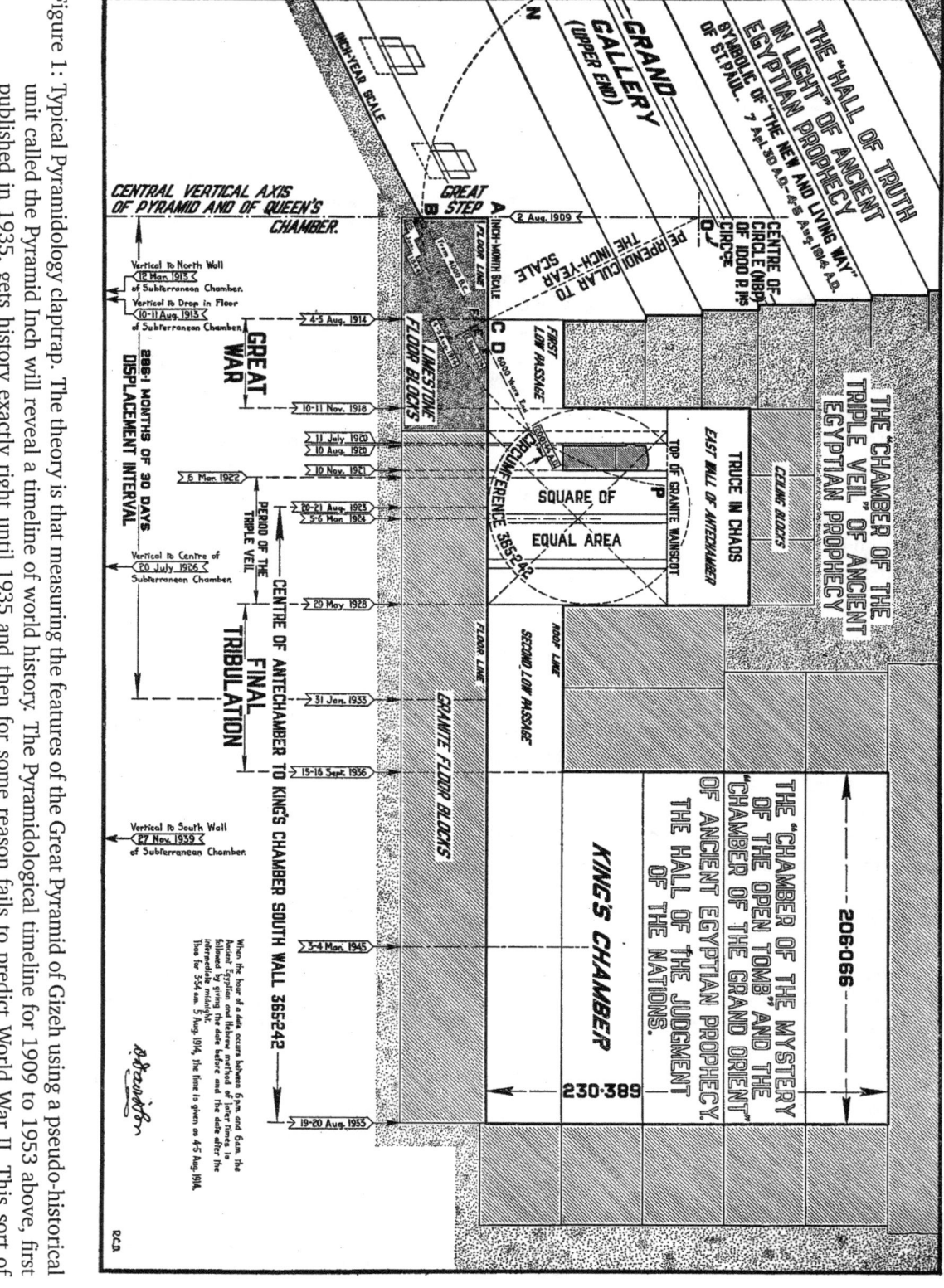

Figure 1: Typical Pyramidology claptrap. The theory is that measuring the features of the Great Pyramid of Gizeh using a pseudo-historical unit called the Pyramid Inch will reveal a timeline of world history. The Pyramidological timeline for 1909 to 1953 above, first published in 1935, gets history exactly right until 1935 and then for some reason fails to predict World War II. This sort of nonsense discouraged serious scholarly work in historical metrology for most of the 20th century

back to ancient originals, but the complete absence of any source information makes it impossible to judge the author's reasoning. A. E. Berriman's 1953 *Historical Metrology*, on the other hand, is scrupulous in its scholarship but vitiated by an underlying theory that really doesn't make sense,[3] even though Berriman arrives at some of the same conclusions that I eventually did. And neither of these efforts focuses on the system we use in the United States.

When I began this inquiry in the early 1980s, I found just three works that could be considered credible sources for a history of the origins of our Customary System. The first was a U.S. government publication from 1821—a masterful report by John Quincy Adams, then Secretary of State, summarizing four years of intensive study of the history of our system of weights and measures undertaken at the request of Congress. The second was the article on ancient weights and measures written for the 11th Edition of the *Encylopædia Britannica* by the legendary Egyptologist W. M. F. Petrie, then the greatest and still one of the leading authorities on the subject. And the third was the 1967 book *Weights and Measures* by F. G. Skinner, who inherited Petrie's enormous hoard of metrological specimens and was for some 30 years the curator of the Weighing and Measuring Collection at the Science Museum in London. These introductions were the impetus for my further investigations, which have proceeded by fits and starts for over a quarter of a century while I was spending most of my time working in the computer industry.

In the interim, we have been blessed with a major scholarly work, *The Weights and Measures of England* by R. D. Connor (1987)[4]—a source with which I do not always agree but which I cite for many of the facts related in this book. But we still don't have a popular yet reasonably well researched history of the Customary System we use in the United States, which the British abandoned almost 200 years ago. And while archaeologists seem to have taken a renewed interest in the subject, they continue to be overly shy about tracing current standards back to their ancient originals, even though the evidence in many cases is stronger than evidence for the reconstruction of Indo-European roots by linguists or the tracing of cultural influences by art historians, to take two closely related examples. In this regard, the amateur historian is at a distinct advantage.

The intended reader of this book is someone who wants to know where the Customary standards we use every day in the U.S. ultimately came from and is willing to follow me in untangling a sometimes confusing history. Since the Customary System in its current form is basically Elizabethan, this book may also prove useful to people who take a special interest in late medieval England, whose culture of measurement we have continued from colonial times virtually unchanged. I assume no previous knowledge of the subject on the part of the reader than I had when I began this inquiry, but I have also made no attempt to simplify it by dumbing down or modernizing the language of my sources.

In light of the miserable state of research left me by most previous works on the subject, I have felt duty bound to cite my sources where possible, resulting in a profusion of footnotes. I beg pardon of the great majority of readers who will find this information largely irrelevant and hope that they will understand my wish to leave the history of our Customary System in better shape than I found it. Professional historians, on the other hand, will note some places where I have failed to cite a source for some assertion or other; these are cases where my notes preserved facts whose original sources were lost at some point in the quarter century that elapsed between my initial research and the production of this book. I can only ask the reader to trust that unsubstantiated claims were once properly attributed and that observations that I believe to be original with me are so noted, or at least clearly implied.

Jon Bosak
Ithaca, New York
August 2010

Notes

[1] From Davidson, *The Hidden Truth in Myth and Ritual and in the Common Culture Pattern of Ancient Metrology*, 59.

[2] By contrast, Pagan use of the Celtic/Saxon measurement tradition appears to lean more to geomancy than eschatology; for example, *Natural Measure* by Nigel Pennick. It's better researched, too.

[3] Berriman believed that the major ancient systems of weights and measures were based on the density of gold, similar to the

NOTES

way that the metric system is based on the density of water. But there is no convincing practical motivation for such a basis.

[4]R. E. Zupko's 1977 book *British Weights & Measures: A History from Antiquity to the Seventeenth Century* is another important contribution, but it focuses mainly on the history of the enforcement of weights and measures regulations in England rather than on the standards themselves.

Acknowledgements

Chief credit for their help with this book must go to my family; primarily, of course, to my wife, Bethany Schroeder, for putting up with this project for 27 years with the patience that only a fellow writer could have mustered; and to our daughter, Clara, who had not been born when the project began but was able to lend scholarly and editorial assistance to its completion.

Next, my thanks go to the helpful staffs at the various libraries I visited in the course of my research, which took place almost entirely in the days before the existence of the World Wide Web: Sterling Library at Yale University; Rivera Library at the University of California, Riverside; Green Library at Stanford University; Trinity College Library, Dublin; the National Library of Ireland; Mann Library at Cornell University; and the law libraries at Cornell University and the University of California, Berkeley. I am especially grateful to the head librarians of Olin Library, Cornell, and the University Library, UC Berkeley, for granting me stack access in an era when the stacks were restricted to accredited researchers. Thanks are also due to the curators of ancient artifacts at the Metropolitan Museum in New York; the British Museum in London; the Louvre in Paris; the Boston Museum of Fine Arts; and the University of Pennsylvania Museum of Archaeology and Anthropology in Philadelphia for maintaining the collections that give amateur historians such as myself the basics of their education and that provided a number of leads to the facts collected in this book.

For their assistance in clarifying details of the current state and recent history of the U.S. Customary System, I am indebted to Philip C. Freije, Deputy Chief Counsel for Economic Affairs, U.S. Department of Commerce, and to Louis E. Barbrow, Consultant, Office of Weights and Measures, U.S. Bureau of Standards (now the National Institute of Standards and Technology). I am particularly grateful to Karma Beal, Historical Information Specialist in the Information Resources and Services Division at NIST, for her great kindness in providing me with a reproduction of the exceedingly rare copy of W. M. F. Petrie's *Inductive Metrology* in the Bureau's library.

For his material contribution to the resolution of a historical puzzle related to the definition of our Customary measures of capacity, I thank David J. Eveleigh, Curator of Agricultural & Social History at the City of Bristol Museum, who provided the measurements of the Henry VII wine gallon presented in Chapter 4.

I would also like to thank the many respondents to the international survey of export wheat and barley test weights I carried out in 1985. Selected results of that survey are presented in Appendix A together with the names of those who provided the data; acknowledgement of the rest will come when I publish a full account of my research into the subject. Special recognition, however, must be extended to Dr. Mark E. Sorrells, Professor of Plant Breeding at Cornell University, whose gracious permission to review decades of Cornell laboratory data allowed the question of grain weight as a natural standard finally to be put upon a sound empirical basis.

Introduction: Postdecimal numbers

To fully understand the history of our customary units of measure, it's necessary to recognize the special significance of certain numeric factors you might not ordinarily notice.

Users of any system of numeration quickly learn that certain numbers are especially significant in that system. Thus, as users of a decimal system, we all recognize powers of 10 as special (a hundred, a thousand, a million) as well as simple multiples of those powers (for example, 20, 400) and binary divisions of those numbers (for example, 500, 250, 125).

Often in the discussion ahead we will encounter numbers that are not significant in the decimal number system but are very significant if you mix in factors of 3. Multiples such as 12, 24, 60, 288, 450, 480, and 5760 look strange to us now, but in the past these numbers were very practical because they gave the merchant, goldsmith, money-changer, and druggist so many different ways to divide things up, or, coming the other way, to assemble them.

Consider the number 60, for example. It can evenly be divided into halves, thirds, quarters, fifths, sixths, tenths, twelfths, twentieths, and thirtieths. This wealth of divisors is due to the fact that 60 contains all three of the lowest prime factors greater than one, i.e., 2, 3, and 5. Purely decimal relationships (10, 100, 1000, etc.) contain only the factors 2 and 5, and thus cannot be evenly divided by 3 and its multiples. They don't contain enough 2s, either, which makes them difficult to work with if halved very far (250, 125, 62.5, 31.25, 15.625, ...) As John Quincy Adams said, "decimal arithmetic, although affording great facilities for the computation of numbers, is not equally well suited for the divisions of material substances."[1]

Now consider the assembly of things in the world rather than their division. Collections of objects frequently arrange themselves physically in forms that don't fit a system based on 5. This is why bottles of beer, for example, are sold by 6 and 12 and 24 rather than 5 and 10 and 20 (think about this).

It's also why a surprising number of these ancient nondecimal factors (often with a couple of zeros attached: 2400, 9600, 14400 or "14.4K" and so on) show up repeatedly in parameters associated with computer hardware. These numbers simply reflect the ways that physical objects can most efficiently be arranged in the real world.

For obvious reasons—look at your hands—10 is the most primitive basis for counting things. The ancient Egyptian system of weights and measures was strongly decimal, and the traditional Chinese system almost exclusively so.[2] The introduction of 3s was a later development that led to a more sophisticated numbering system, one based on the number 60 rather than the number 10. This became the system of science, in particular of astronomy, and it lives on in our division of the hour and minute by 60, the division of the circle by 360, and the division of the difference between the freezing and boiling points of water in the Fahrenheit scale by 180. Thus the German language, for example, still has the word *Schock* "group of 60" and the verb *schocken* "to count by 60s." In medieval England, the word "hundred" often meant 120, not 100. The Sumerian system was so thoroughly based on 60 that its users did not even have words for 100 or 1000; for the former, they would say "60 and 40," and for the latter, "16 60s and 40."[3]

The sexagesimal (base-60) system of the Babylonians and Sumerians used a positional notation not unlike our own. Instead of representing powers of 10—that is, 1s, 10s, 100s, 1000s, etc.—each position represented a power of 60: 1s, 60s, 3600s, etc.[4] Modern archaeologists represent these numbers by writing the "digits" right to left, just as we do for decimal digits, but separated by commas, with the Babylonian number in each place represented by an integer from 0 to 59. So, for example, the sexagesimal number 3,4,5 is $(3 \times 3600) + (4 \times 60) + (5 \times 1)$ or 11,045 in decimal notation.

The same system served for fractions, just as in our decimal system (and indeed, it was for fractions, that is, divisions, that it survived in the modern usages cited above). Thus we find in an Old Babylonian table of reciprocals[5] the value for 27 (i.e., $\frac{1}{27}$) given as 2,13,20, that is, $(2 \times \frac{1}{60}) + (13 \times \frac{1}{3600}) + (20 \times \frac{1}{216000}) = 0.037037037\ldots$

Notice that the sexagesimal system allows the simple non-repeating expression of this fraction in a way that we cannot duplicate in decimal notation. Of the first 10 integers, only one (7) has a reciprocal that can't be represented by a non-repeating sexagesimal expression.

In everyday use, the value of the unit was generally known from the working context and was scaled so that every working value could be represented by just three digits no matter what the absolute size of the quantity. Each triple from 0,0,1 to 59,59,59 in the Babylonian notation represented a value between 1 and 215,999 (i.e., (59×3600) + (59×60) + 59). In other words, the base-60 triple provided something rather better than five decimal places of precision, which is sufficient for the great majority of practical real-world applications even now.

The old sexagesimal system is important in what follows because "round numbers" in this notation correspond to factors we will see repeatedly as we go; for example:

0,1,0	60	0,12,0	720
0,2,0	120	0,15,0	900
0,3,0	180	0,16,0	960
0,4,0	240	2,0,0	7200
0,5,0	300	4,0,0	14400
0,6,0	360	10,0,0	36000
0,8,0	480	12,0,0	43200
0,9,0	540	15,0,0	54000
0,10,0	600	16,0,0	57600

Since the primitive decimal basis remained a strong influence, we will often encounter these numbers divided or multiplied by 10 or 100.

In the Greco-Roman tradition that gave us our Customary System, dozens (12s) play a strong role, base-12 being a simplified or degenerate form of base-60 with its own series of multiples, the most common being 24, 36, 48, 60, 72, 96, and 144.

The Greco-Roman tradition also had a strong binary component, and divisions into 12 were often used alongside a parallel system of divisions into 16, the Romans leaning toward 12 and the Greeks tending more to 16. The Greek preference—and, as we shall see, certain inherent practical advantages in weighing and the measurement of capacities—gave numbers in the binary sequence (2, 4, 8, 16, 32, 64, 128) special status as well.

Most premetric European systems of length, weight, and capacity related their units by factors roughly balanced between decimal, base-12, and binary sequences. Our Customary System is the last survivor of this tradition.

Notes

[1] ADAMS, 70. See "Frequently Referenced Sources" on page 151.

[2] The use of decimal notation by the Chinese goes back to the 14th century B.C. Decimal division of Chinese length units goes back to at least 400 B.C. (Needham, *Science and Civilization in China*, 3:84–89).

[3] Thureau-Dangin, "Numération et Métrologie Sumériennes," 125. For an in-depth treatment of the origins and later survivals of base-60 numeration, see Thureau-Dangin, "Sketch of a History of the Sexagesimal System."

[4] Neugebauer and Sachs, *Mathematical Cuneiform Texts*, 2.

[5] Ibid., 11. The system was supplemented by special symbols for the fractions $\frac{1}{3}$, $\frac{1}{2}$, $\frac{2}{3}$, and $\frac{5}{6}$.

Chapter 1

The Northern foot

Standard measures have great persistance compared with most other cultural artifacts. Land measures tend to be particularly stable, frozen as they are in features of the constructed landscape and often defended by economic interests. We will, for example, be stuck with our existing U.S. Customary land measures for centuries to come.

Such a system should be logical and easy to work with, and so it is. But one would never know this from the way it has been explained over the past several hundred years.

As schoolchildren, we learned our mile at first as a blessed mystery, and then as an object of derision, to be trotted out whenever someone needed an easy proof of the decrepitude of the Customary System: one mile is... 5280 feet. How grotesque. But this is the wrong way to view our land measures. Let's take another look, remembering that we are dealing with the local variety of a system almost as old as the Indo-European languages and developed originally by farmers.

Customary land measure

Horizontal and vertical measurements customarily use different units. Thus in modern times, we say 10 miles horizontally but 50,000 feet vertically, 16 kilometers horizontally but 16,000 meters vertically. In ancient times, volumes the size of buildings and granaries were measured in feet horizontally, for obvious reasons, and in cubits vertically, because a cubit was originally the distance from the elbow to the tip of the middle finger and is the easy module with which to measure how tall something is.

Most premetric systems of measurement had a unit somewhere between 11 and 14 inches long that historians generically call a foot. Just as with our foot, however, these units did not correspond to real human feet, which average 9 to 10 inches in length, and often the word used to refer to the unit didn't actually mean "foot" at all but rather something like "unit" or "measure."

In a minority of premetric systems of length (most notably the traditional system of the Spanish-speaking countries) there was a unit, called by metrologists a "natural foot," that was about the length of an average adult foot, but in the majority of systems, at least in the western tradition, the unit we call a foot was defined as two-thirds of a cubit.

The earliest land measure was probably the pace, usually divided into five feet (as it still is in the U.S. Customary System), and the half pace, or step, usually $2\frac{1}{2}$ feet. In the Customary System, these units are called the geometric pace and the military pace. The ancient Romans called them *passus* and *gradus*. In the minority of systems that use the shorter natural foot, the pace is divided into six natural feet and the half pace into three.

The basic unit of land measure in our current U.S. Customary System is the rod,[1] which was originally three paces but is defined nowadays as $16\frac{1}{2}$ feet. The later definition is clearly insane, but let's let that go for the moment and just consider the rod itself. The rod is the length of a long pole (another traditional name for the unit) just about the length of a pole vaulter's pole—that is, just as long as it can be and still be carried around easily when it's necessary to lay out a plot of land or decide whose piece of land a plot belongs to. A system that works well for laying out gardens and small fields—areas about the size that one man with a plow can manage—looks like this:

4 rods make a chain

10 chains make a furlong

Chapter 1. The Northern foot

This is the U.S. Customary System of lengths for land measure, and the units are still embedded in many older land titles in this country.

In the oldest English statutes, the rod is called a *perch* (Latin *pertica*), and the term *pole* is also frequently encountered.[2]

A furlong is a "furrow long." The furlong and the rod are farmer's measures, whereas the chain is a surveyor's measure.

To put it in terms more familiar to most people today, a furlong is about the length of a long city block. Implicitly we still use the furlong when we use the quarter mile to measure things like distances to freeway offramps, because a quarter mile is really just a double furlong; a furlong is exactly an eighth of a mile. In other words:

1 mile = 8 furlongs = 80 chains = 320 rods

The mile (originally a thousand paces, in Latin *mille passus*) goes back to the Romans, but of course their mile was based on the Roman foot, which was a little shorter than ours. The oldest legal definition of the Customary mile (as 8 furlongs) is given very late, almost in passing, in the English statute 35 Elizabeth 1 c. 6 §8 (1593)[3] limiting the construction of new buildings in the City of London. The mile was not created with that statute, however; the first edition of Holinshed's *Chronicles* (1577) presents the relationship 8 furlongs = 1 mile as traditional. The MED (see "Frequently referenced sources" on page 151) quotes the same definition from a manuscript of 1500 and traces the word in written English back to 1125.

The older statute references to the mile are obscured by the fact that the English word *mile* does not appear in Norman French, in which all the English laws were written until the end of the 15th century; the word *mile* in Norman French of the time simply meant "a thousand," and the laws written in French represented English miles with a term that is variously spelled "lewes," "leues," "leukes," "leukez," or "leuges," as for example in 28 Edward 3 c. 14 (1354) and several later statutes. The word is obviously cognate to English *leagues,* but it is used throughout the old Anglo-Norman statutes to mean English miles, the same as ours today. The earliest statute mention of the English word *myles* occurs in 19 Henry 7 c. 1 (1503), shortly after the laws began to be written in English, and there seems to be no reason to believe that the unit (whatever it was called) is not as old as the rest of the land measures.

Customary area measure

Our Customary System of area measure works like this (Figure 1.1):

1 perch (for gardens) = 1 square rod

1 acre = a rectangle 4 rods (1 chain) in width and 40 rods (1 furlong) in length = 160 perches

1 square furlong = 10 acres = 100 square chains

Notice that the acre was not originally defined as a square but as a narrow strip of land 4 rods (1 chain) in width by 40 rods (1 furlong) in length, just big enough to support the family of a subsistence farmer. This, we are told, reflects Anglo-Saxon land ownership practices, each farmer's cottage being located at the end of his holding. Ten of these strips laid side by side make a square furlong. A square mile is 8 furlongs on a side and contains 64 square furlongs or 640 acres (Figure 1.2).

The measure of 4 rods—the width of, among other things, the standard cricket pitch—was known for centuries simply as the acre-breadth; it didn't get the name *chain* until surveyors rationalized the acre into decimal divisions in the 17th century.

If you take acre-rectangles 4 rods by 40 and lay them end to end instead of side by side you get a highway whose surface occupies 1 acre to the furlong or 8 acres to the mile. This is literally true where I live; by a New York colony statute of 1703, the first system of state highways was laid out with a standard minimum width of "the Breadth of Four Rod of *English* Measure,"[4] and the old acre-breadth remains the width of the right-of-way that the State of New York claims for the highway that runs by my door today.

A quarter of an acre is called a rood. It, too, was visualized as a strip the length of a furrow, in this case just 1 rod wide by 40 long instead of 4 rods wide by 40 long in the case of the acre.[5] Both the rood and the acre are, as these oldest definitions suggest, designed to make it easy to estimate the time it would take (or the number of teams you would need) to plow a given area of land.[6] Most traditional units of land area were either defined in terms of the amount of time it would take to plow them or were given the name of the capacity measure containing the amount of seed it would take to plant them; our acre belongs to the first of these traditions.

The rood used to be quite a common land measure. Given the frequency with which "quarter acres" are seen in advertisements for land in the U.S., we could say that it still is a common unit, though we have forgotten its proper name.

The relationship

10 acres = 1 square furlong

makes areas expressed in acres fairly easy to visualize if you're used to distances in miles. Two hundred acres, for example, is equal to 20 square furlongs, suggesting to the mind's eye a holding measuring 4 furlongs by 5 furlongs or $\frac{1}{2}$ mile by $\frac{5}{8}$ mile in extent. Forty acres (as in "40 acres and a mule") is a square $\frac{1}{4}$ mile on a side, and 160 acres is a square $\frac{1}{2}$ mile on a side.

The larger units used to have their own names: a square furlong was known in Anglo-Saxon as the ferlingate or ferdelh; 4 ferlingata (i.e., 40 acres) made a virgate,[7] and 4 virgata (160 acres) made a hide.[8] The square furlong was often just called a furlong (area being understood from the context), and with this meaning, the word was frequently an element in the names of fields (for example, in a land register from 1500, *Fulewelfurlonge, Chaldewelfurlonge, Schortfurlonge, Brocforlonge, Middelfurlonge*).[9]

Modern equivalents

The absolute sizes of our acre, rod, and furlong have remained constant for at least 1300 years, and probably much longer.[10] This ancient system, which underlies the surveys upon which land ownership is based in the United States, is not only perfectly comprehensible but rather charming. The craziness comes in when these simply related units are defined in terms of modern feet:

1 rod = $16\frac{1}{2}$ feet

1 chain = 66 feet (22 yards)

1 furlong = 660 feet (220 yards)

1 mile = 5280 feet

The area definitions are even worse:

1 perch = $272\frac{1}{4}$ square feet

1 acre = 43,560 square feet

After noting the underlying simplicity of the basic system of land measures, a commission of the British Parliament appointed in 1841 sadly observed that

> the length of the land-chain (22 yards) is practically inconvenient when used for any other measures than those related to the mile and the acre; and, theoretically, it is the greatest anomaly in the whole English system of measures and weights. The length of the mile also, as expressed in the measures best known to the greatest part of the English people (the yard and the foot), is given by the most inconvenient numbers. . . . [11]

The strange and inconvenient relationship between the longer land measures (rod, chain, furlong, mile) and the shorter units of everyday use (inch, foot, yard) can be traced back to the time of Edward I (1272–1307). A law titled *Compositio Ulnarum et Perticarum* (Composition of Yards and Perches), of uncertain date but traditionally given the citation 33 Edward 1 stat. 6 (1301),[12] reads as follows:

> *Ordinatum est quod tria grana ordei sicca & rotunda faciunt pollicem; duodecim pollices faciunt pedem; tres pedes faciunt ulnam; quinque ulne et dimidia faciunt perticam; & quadrigenta pertice in longitudine, & quatuor in latitudine faciunt unum Acram.*

That is,

> It is ordained that three grains of barley dry and round [laid lengthwise] do make an inch;

Chapter 1. The Northern foot

☐ A *perch* is a square *rod* (1 rod = 16.5 English feet or 15 Northern feet). A perch is the size of a small garden plot or the area occupied by a small fruit tree.

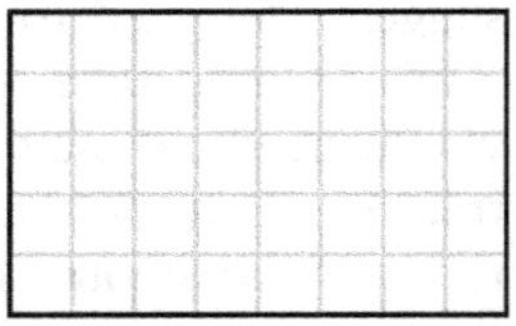

40 perches make a *rood*. This is about the size of an orchard. But that's not the way the rood was originally defined.

Conceptually, the rood is a long, skinny strip 1 rod wide and 40 rods long.

The rood is a plowman's measure. The length of 40 rods is a furrow long — a *furlong*. Each rood represents eight furrows, side by side.

Four roods make an area 4 rods by 40 rods called an *acre*. An acre contains 160 perches. The width of the acre (4 rods) is a surveyor's unit, the *chain*. So the acre can also be described as the area measuring 1 chain by 1 furlong. It's the area of land that one man with one team can plow in one day — 32 passes with the plow.

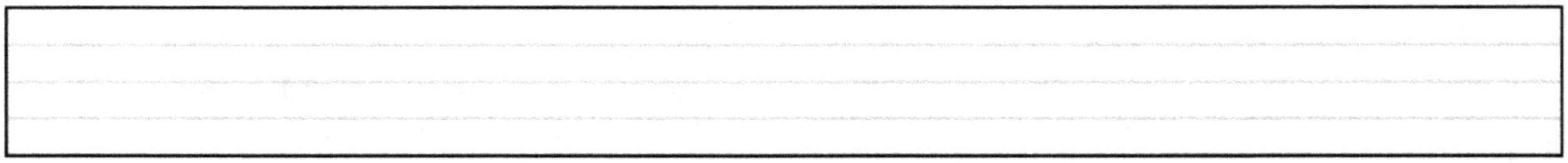

Figure 1.1: Basic Customary area measurement

twelve inches make a foot; three feet make a yard; five yards and a half make a rod; and forty rods in length and four in breadth make an acre.

And this is the law that defines these measures in the U.S. to this day.[13]

It's important to understand that the statute doesn't actually define the rod and the acre in terms of the inch, foot, and yard, though it seems to be doing that in this passage; rather, it defines the inch, foot, and yard by their relationship with the much older rod and acre. The unwieldy set of relationships between the land measures and the foot/yard system is explained by one simple historical fact: the modern foot of our system has nothing to do with our land measures. The land measures are based on a different foot to which our modern foot is not directly related.

The Northern foot

The ancient foot of our land system is one of the most widely distributed of all premetric measures. It can be found either explicitly or implicitly in the premetric systems of western Europe, Russia, and parts of China, as well as anciently in Syria (at 13.18 inches in 900 B.C. and 13.22 inches in A.D. 620) and in other places at outlying values ranging from 13.1 to 13.3 inches. Notice that I'm not just pointing to the existence of a notionally foot-like thing; I'm talking about a specific standard of length.

Because of its northern distribution in historic times, this old unit was given the name "Northern foot" by Flinders Petrie. According to Petrie and other historians, the ancient foot standard (at 13.2 inches exactly) remains embedded in our traditional land measures like an insect in amber:

Chapter 1. The Northern foot

Here's the acre seen from higher up.

An area of ten acres is a *square furlong*. You can also think of a square furlong as an area 40 rods by 40 rods containing 40 roods.

Eight furlongs make a *mile,* so there are 64 square furlongs (640 acres) in a square mile.

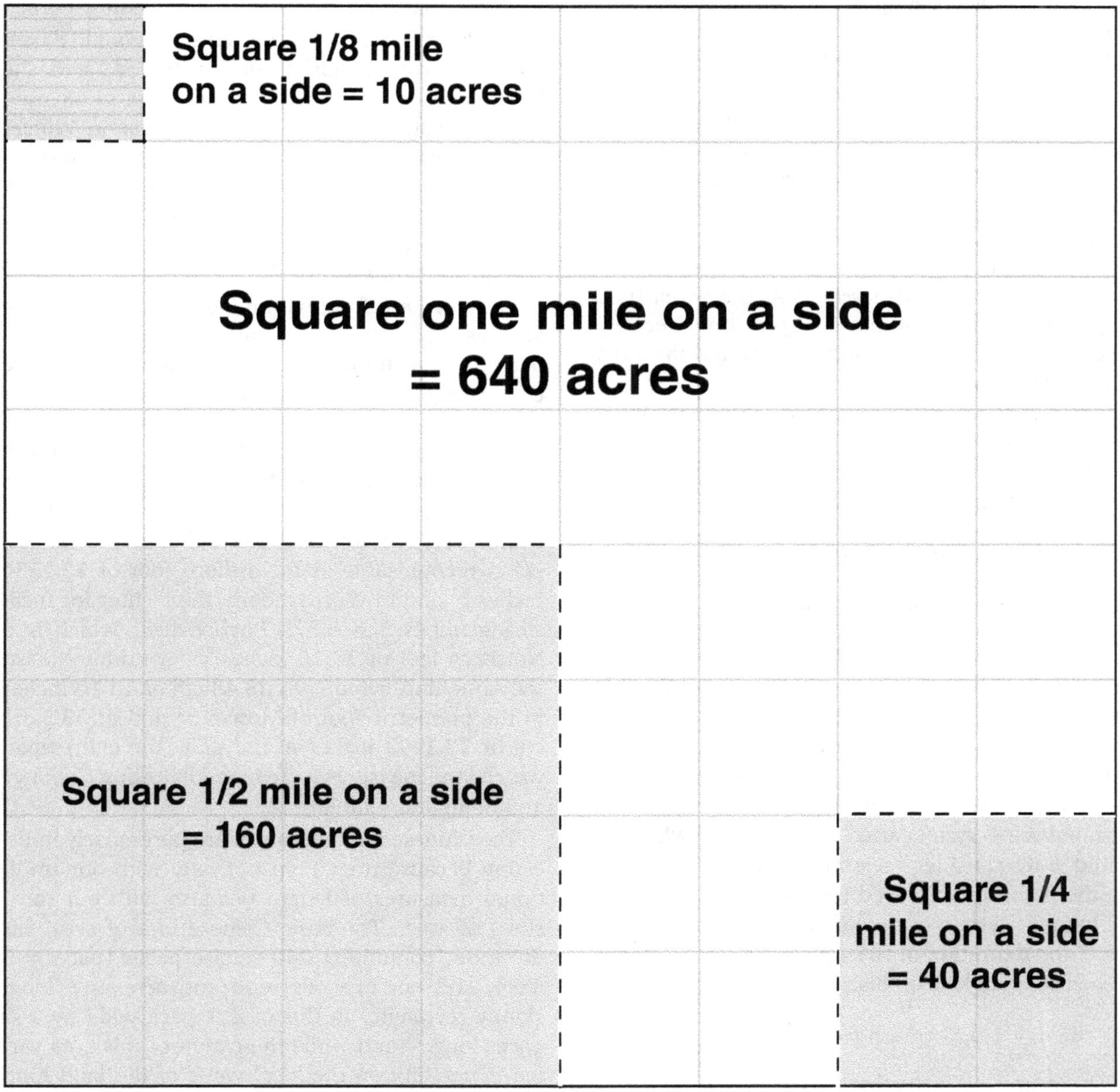

Square 1/8 mile on a side = 10 acres

Square one mile on a side = 640 acres

Square 1/2 mile on a side = 160 acres

Square 1/4 mile on a side = 40 acres

Figure 1.2: Larger land areas

Chapter 1. The Northern foot

1 rod = 15 Northern feet

1 chain = 60 Northern feet

1 furlong = 600 Northern feet

1 mile = 4800 Northern feet

To these units we should add a Northern pace of 5 Northern feet, evidence for which has been found in Saxon villages in England.[14] Our mile is 960 of these paces.

A very late reference to the Northern foot appears in a manuscript from the time of Edward I (in whose reign was enacted the 1301 statute quoted above) that states "12 inches make a foot; and 3 feet make an *aune;* and 5 *aunes* make a perch; and 4 perches in width and 40 in length make 1 acre of land" *(xii pouces font 1 pie; et trois piez font l'aune; et v aunes font une perche; et iiii perches en laeure et xl in longure font 1 acre de terre).*[15] As the perch (i.e., rod) and the acre have not changed, this precisely specifies the foot as the Northern foot and the *aune* as the Northern yard.

Stated in terms of the foot upon which they were originally based, our customary units look positively reasonable. The area measures are seen to be equally simple:

1 perch = 225 square Northern feet (15×15)

1 acre = 36,000 square Northern feet

Compare these definitions with the $272\frac{1}{4}$ square English feet for the perch and 43,560 square English feet for the acre that we've been trying to work with using an unrelated set of units since 1301.

The simple Northern-foot-based definitions look even better if expressed in terms of the base-60 numbering system used anciently for what we would now consider "scientific" calculation. Using the notation employed by scholars to represent the base-60 system of the Sumerians and Babylonians,[16] the definition of the acre as the area 1 chain by 1 furlong looks like this:

1 chain = 1,0 Northern feet

1 furlong = 10,0 Northern feet

1 acre = 10,0,0 square Northern feet

The Sumerians attached so much importance to the number 10,0,0 (36,000) that they had a special name for it, *šar-u*, and a special symbol, ⊚.[17] It's also notable that the furlong is defined in terms of the Northern foot (600 to the furlong) exactly the way the ancient Greek equivalent was defined, the Greek stadion being 600 Greek feet. So properly understood, the ancient system underlying our land measures is not only sensible, it is also in line with a tradition that shaped European culture in general.

The Northern foot is very old in England. Saxon villages have already been mentioned, and the Northern foot appears to have been the unit used in the construction of the Old Minster in Winchester, which was built in A.D. 648.[18] At an average value of 13.22 inches, the Northern foot was more commonly used than the modern foot in old English buildings, and its use for that purpose is found as late as the 15th century.[19]

The Northern foot, with minor variations, can be seen as the basis for a number of premetric European and Asian units of length whose origins were completely unknown to their latter-day users. For example, the premetric French canne of Marseilles, a land unit measuring 79.238 inches, is 6 Northern feet of 13.206 inches.[20] A thousand square cannes of Marseilles is thus one English acre.[21] In the 19th-century Tyrol, the fuss or piede tirolese of 33.4097 cm is recognizable as a Northern foot of 13.1534 inches,[22] and in Württemberg, the lachter for measuring mines was 6.576 English feet, which is 6 Northern feet of 13.15 inches.[23] Continuing east, the same unit is found at 33.43 cm or 13.16 inches in the premetric foot of Moscow[24] and at 33.5261 cm or 13.1993 inches in the unit (the chih) upon which land measures were based in Peking (Beijing) and Shanghai, China.[25]

The Chinese land measures are particularly interesting because they align not only with our traditional measures of length but also with our measures of area. The basic Chinese unit of area was the mow, containing 240 square paces (pace = 5 feet), laid out like our rood and acre as a long, skinny rectangle, in this case 1 pace wide by 240 paces long;[26] and while the extent of this area varied depending on the local value of the land foot, the figures just given for Peking and Shanghai show that in these centers the mow was 1 Northern pace

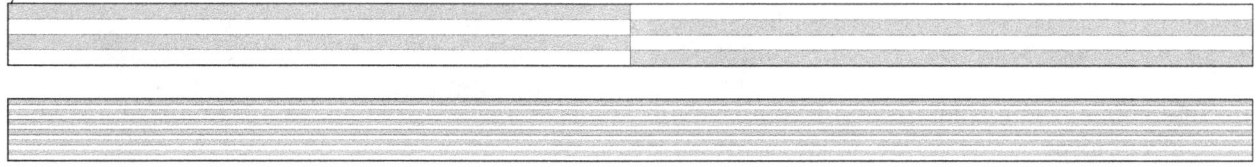

Figure 1.3: Two English acres, divided into 8 English roods (top) and 12 Chinese mows (bottom)

wide by 2 Customary (Northern) furlongs in length. If you compare the mow of Peking and Shanghai with the English rood and acre you'll see that

3 Chinese mows = 2 English roods

and

6 Chinese mows = 1 English acre

(see Figure 1.3).

A unit of 2 Northern feet (the Northern cubit) survived into the 19th century in Europe and North Africa as a widely employed wool and cloth measure under the names ell, aune, covado, braccio, or pic, depending on the location. An official survey undertaken by Lord Castlereagh in 1818 recorded values in 28 different cities that can be summarized as follows:[27]

Germany:
    avg. 26.24 in. (2×13.12 in.)

Holland and Belgium:
    avg. 26.54 in. (2×13.27 in.)

France and Portugal:
    avg. 26.66 in. (2×13.33 in.)

Italy and the Adriatic:
    avg. 26.63 in. (2×13.31 in.)

North Africa, Syria, Turkey, &c.:
    avg. 26.76 in. (2×13.38 in.)

From this it can be seen that the Northern cubit used as a cloth measure had a value closest to our version of the standard in the countries farthest north and became increasingly larger as it moved south.

The Northern pace (five Northern feet or 66 inches) seems to have been even more widely distributed than the Northern foot itself, appearing in a number of traditional systems divided into six feet (the "natural feet" noted earlier) instead of five. The model for this alternative division was the vara of Spain, which at 83.5905 cm (32.9096 inches) in Madrid was half a Northern pace, corresponding to $2\frac{1}{2}$ Northern feet of 13.1639 inches. In the *General Laws of the Territory of Colorado* (Third Session, 1864), the vara "in so much of this Territory as was taken from New Mexico" was defined as equal to 33 inches, which is $2\frac{1}{2}$ Northern feet of 13.2 inches exactly.[28] All the Spanish and Mexican land grants in California were reckoned using a vara of 33 inches as well.[29] In the Spanish tradition, the vara was divided into three feet of about 11 inches. The Northern pace survived into the mid-20th century as the ragil of Anglo-Egyptian Sudan, at 1.833 (i.e., $1\frac{5}{6}$) yards[30] the English version of the Northern pace (66 inches) exactly, and the half pace was still in use as the gaz memar of Afghanistan, which at 83.6 cm[31] was virtually identical to the Castilian vara. The Northern pace also appears as the klafter of Heidelberg, 6 "natural feet" of 27.85 cm[32] or a total of 65.79 inches, that is, five Northern feet of 13.16 inches.

This alternative division of the Northern pace and Northern rod (3 paces) into shorter "natural feet" as in the Spanish tradition appears to have been employed in Anglo-Saxon building practices as well. Some of the buildings show the rod divided not just into thirds (i.e., paces) but also into sixths, giving measurements such as $\frac{5}{6}$ Northern rod that do not equate to whole numbers of Northern feet but do equate to whole numbers of the shorter foot. This foot is found in Yeavering at 281 mm, and a statistical analysis yields a foot of 280 mm at Jarrow;[33] these shorter feet correspond to a Northern pace of 66.14–66.38 inches and a Northern foot of 13.23–13.28 inches.

Sometimes we even find the shorter Northern foot divorced from the pace, as in the case of the

Chapter 1. The Northern foot

palma of Moldavia (modern Romania). The palma was 8 palmacs of 3.486 cm = 27.888 cm (10.980 inches);[34] three of these would make a vara (half Northern pace) of 83.664 cm or 32.939 inches ($2\frac{1}{2} \times 13.175$ inches), though the Moldavian system did not organize them this way, instead defining the stanjen as 8 palmas or 223.104 cm (87.836 inches).[35] The builder's canna of Rome, 223.425 cm, was practically identical to the Moldavian stanjen, but in Rome it was divided into 10 palmi that would each be $\frac{2}{3}$ of a Northern foot of 13.194 inches.[36]

Earlier I noted that our mile, which is defined anomalously as 5280 feet, is actually 4800 Northern feet. Northern measure also explains at least three other European "miles" that otherwise bear no sensible relationship to the rest of their national systems. The old meile of Brunswick, Germany, was 34,124 Rhine feet;[37] this is 32,000 Northern feet of 13.177 inches. In Hungary, the merföld or meile was 8353.6 meters, which bears no whole-number relationship to any of the other Hungarian units of measure, but is 25,000 Northern feet of 13.155 inches.[38] The 19th century meile of Lithuania was 8.039816 km; it can't be divided into the surviving shorter units (which were all Russian), but it can be divided into 24,000 Northern feet of 13.19 inches.[39]

Archaeology of the Northern foot

The Northern foot exhibits three noteworthy features:

- Its range of distribution, from England in the west to China in the east. Petrie called the Northern foot "the most widespread [premetric] standard that we know."

- Its relative stability: the most important independently surviving versions of the Northern Foot in 19th century Europe, Russia, and Asia show a variation of less than one percent from the value implied by our current U.S. land measures.

- The maintenance of each local variation as a separate standard, demonstrating that all memory of a common origin had been lost by the 19th century.

Figure 1.4: Fragment of a ruler from Mohenjodaro (2500 to 1500 B.C.)

Figure 1.5: Reconstruction of the ruler fragment

Taken together, these characteristics suggest a standard of great antiquity rooted in a major ancient civilization, and the archaeological data, while scant, are enough to confirm this. The parent of the Northern foot is found in the cradle of the Indus Valley civilization at the famous site known as Mohenjodaro (2500 to 1500 B.C.), where 20th century archaeologists found the accurately divided fragment of a ruler engraved on shell that you see in Figure 1.4.[40]

As its discoverers noted, shell is probably the best material of which the ruler could have been made, being unlikely to warp or crack. The fragment "is beautifully made and finished," the division lines "very carefully cut with a thin saw" and averaging 0.02 inches in width and depth. The five divisions shown here average 0.264 inches apiece with an average error of just 0.003 inches, and there can be no doubt that the object was part of a measuring device. The fact that there are five divisions between the dot-marked line and the circle-marked line strongly suggests a decimal system of length measurement, and the standard to which the system belongs is easily identified; the distance between the dot mark and the circle mark is 1.320 inches, or $\frac{1}{10}$ of a Northern foot of 13.20 inches.

It's notable that the same foot is found both east and west of the Indus Valley. Divided decimally, it lines up perfectly with the Peking land chih of 13.199 inches, which was divided into 10 subunits called t'sun of 1.3199 inches; the unit shown by the fragment found at Mohenjodaro was just what the premetric Chinese on the standard of Peking and

Shanghai would have called a t'sun for land measure. To the west, and roughly contemporaneously, the Northern foot is found "on finely inscribed Egyptian Royal Cubits of the XVIIIth Dynasty, BC 1567–1320" marked at or close to the Egyptian 18th digit, giving a foot of 13.27 in.[41]

Thus, the Northern standard seems to have spread outward from the Indus culture like the original language of the Indo-Europeans, but (driven by trade relationships rather than kinship or migration) it did not spread in all of the same directions. Future research may someday succeed in elucidating a process reminiscent of the spread of language but apparently quite independent of it.

A meter by any other name

The Northern foot is far from unique in its survival into the modern era; many of the world's premetric standards of length can similarly be traced back to a few ancient originals that were carefully maintained in separate places after the fall of whichever empire had established them.[42] The major exception to this general rule is our Customary standard of length—the English yard and its subdivisions. The yard was established, tradition has it, by King Henry I of England, and while its division into feet and inches inherits a common Greco-Roman system of organization, the English yard standard appears to have no clear historical precedent, coming into existence some time between the 10th and 12th centuries.[43] I'll offer some possible theories for the origin of the English inch, foot, and yard in Chapter 6. What is clear is that their ascendance over the old Northern foot and yard as the definitional basis of our land measures had a catastrophic effect on the coherence of the whole Customary System.

The loss of the Northern standard is especially sad given the adoption throughout the rest of the world of a unit—the meter—that is virtually identical to the Northern yard (three Northern feet). The Northern yard embedded in our customary system measures 39.6 inches or 100.584 centimeters, a difference from the meter of less than one percent. If the Northern yard had been taken as the basis of the meter, then the following relationships would be exact:

A kilometer would contain 200 rods

A mile would be 1.6 kilometers in length

A rood (quarter acre) would equal 10 ares (0.1 hectares)

A square furlong would equal 4 hectares

A hectare would contain $2\frac{1}{2}$ acres

As it is, these equivalences, while not exact, are still close enough for most purposes of estimation; a hectare, for example, is actually about 2.47 acres.

It might be objected that adopting a meter of 39.6 English inches would have broken the relationship with a natural constant that the inventors of the meter were trying for (4 million meters = the earth's circumference), but that relationship isn't exact anyway, and in practice the approximation is no more useful than it would have been if the new system had included a simple relationship with the British standard.

Since we are stuck with our land measures for the indefinite future, we might as well make the best of them. I think we should teach our children the relationships between the mile, furlong, chain, rod, perch, rood, and acre—all of which are very easily explained with a few uncomplicated diagrams and form a coherent system that is not only simple to understand but actually happens to be historically true—and abandon the modern foot and its multiples and submultiples by just leaving them out of the tables of equivalents. No one should have to remember that a mile is 5280 feet or that an acre is 43,560 square feet. They weren't defined that way in the first place, and the fact of the matter is that they are not really defined that way now.

Survey foot vs. international foot

Unnoticed by most people, the mile and the rest of our statute land measures lost their legal basis in the foot and yard half a century ago. The story is told in U.S. National Bureau of Standards Letter Circular 1035 (1985):

From 1893 until 1959, the yard was defined as being equal exactly to $\frac{3600}{3937}$ meter. In 1959 a small change was made in the definition of the yard to resolve discrepencies both in this country and abroad. Since 1959 the yard is defined as being equal exactly to 0.9144 meter; the new yard is shorter than the old yard by exactly two parts in a million. At the same time it was decided that any data expressed in feet derived from geodetic surveys within the U.S. would continue to bear the relationship as defined in 1893 (one foot equals $\frac{1200}{3937}$ meter). This foot is called the U.S. Survey foot, while the foot defined in 1959 is called the international foot.[44]

In other words, relationships such as "one mile equals 5280 feet" are no longer exactly correct if by "foot" we mean the foot that we use in making rulers and machine tools, the foot that's 12 inches of exactly 2.54 cm each. The 12-inch-long foot is not the one of which 660 make a furlong, and the 144-square-inch square foot is not the one of which 43,560 make an acre. So these definitions of the last few centuries are now all slightly bogus anyway, and I think we would be better off by simply leaving them out of the picture.

This is basically what surveyors in the Customary tradition ended up doing three centuries ago. Faced with the contradiction between the longer measures (chains and furlongs) and the shorter ones (feet and yards), they gave up thinking of the chain in terms of feet, and in accord with a system first described in 1624 by the English mathematician Edmund Gunter, divided the measure of 4 rods into 100 synthetic units called links. An acre is 100,000 square links. At 20.1168 cm (approximately), the surveyor's link is similar to a double decimeter.

I don't think it's necessary for most people living with the traditional system to think of it in terms of "survey feet" and "survey yards" at all. I believe we should think of the Customary land measures as based upon the metric system. Think of a rod as 5 meters, a chain as 20 meters, a furlong as 200 meters, and a mile as 1600 meters. Think of an acre as an area 20 meters by 200 meters, equivalent to 0.4 hectares (or of a rood as 10 ares, which is prettier). It all makes perfect sense if you just think of the meter here as not quite the meter used for everything else, but rather a slight variation that occurs only in land measurement and isn't different enough from the regular meter to concern anyone but a specialist. After all, with our distinction between the "international foot" and the "survey foot," that's what we do now.

Notes

[1] Etymologies for the names of all of the Customary units will be found in the Glossary beginning on page 139.

[2] See the usage note under "rod" in the Glossary on page 146.

[3] "35 Elizabeth 1 c. 6 §8 (1593)" means "the eighth section of the sixth chapter of the statute passed in the thirty-fifth year of the reign of Queen Elizabeth I, which was 1593." See Appendix B for more on the subject of these statutes as they apply to U.S. law.

[4] *Acts of Assembly, Passed in the Province of New-York* (1719), 65ff.

[5] CONNOR, 37.

[6] Thus the Wycliffe Bible (ca. 1400) translates the unit of land in 1 Samuel 14:14 as "an akir that a peyre of oxen in a day is wont to ere" (*ere* as in *arable*). MED, s.v. "Akir." The key to short form references such as "MED" is provided in "Frequently Referenced Sources" on page 151.

[7] From Latin *virga*, "rod." In practice, the virgate (like many of the medieval measures of area) varied according to the productivity of the land; the canonical 40 acres is the value used in the Domesday Book. See OED, s.v. "Virgate."

[8] SKINNER, 92.

[9] MED, s.v. "Furlong."

[10] CONNOR, 42.

[11] *Report of the Commissioners appointed to consider the Steps to be taken for Restoration of the Standards of Weight and Measure* (1842), 7. The commissioners recommended that the mile be replaced by units of 1000 yards called milyards, but this idea never caught on.

[12] That is, the sixth statute enacted in the thirty-third year of the reign of Edward the First, which was 1301.

[13] See Appendix B.

[14] Huggins, "Excavation of Belgic and Romano-British Farm," 64–65; and "Yeavering Measurements: An Alternative View," 150–52.

[15] Hall and Nicholas, *Tracts and Table Books*, 7.

[16] See Introduction. As far as I know, this way of looking at the Northern foot equivalents has not appeared before.

[17] Thureau-Dangin, "Sketch of a History of the Sexagesimal System," 105).

[18] CONNOR, 46.

[19] Petrie, *Inductive Metrology*, 107; Petrie, *Ancient Weights and Measures*, 41.

[20] SKINNER, 42.

[21] Nicholson, *Men and Measures*, 257–58.

[22] MARTINI, 93 and 259. Note that 19th century compilations of world weights and measures like MARTINI, TATE'S, and JOHNSON'S cited throughout this book had nothing to say about the origins of the premetric standards but assumed them to be entirely arbitrary.

[23] Brough, *A Treatise on Mine-Surveying*, 8.

[24] MARTINI, 389.

[25] MARTINI, 513; this is the value for the land measure of Peking set by treaty between China and Italy in 1866. Hoang, writing in 1897 (*Notions Techniques,* 58) puts the kong, the "mesure traditionnelle juridiquement admise dans le territoire di Chang-hai [Shanghai]," at a metric value of 1.673 m and an English value of 66 inches; this would be 5 Northern feet of either 13.17 inches or 13.2 inches. The *Encyclopædia Britannica* (11th Edition, s.v. "Weights and Measures") puts the "treaty kung" at 78.96 inches, which is 6 Northern feet of 13.16 inches (compare with the canne of Marseilles cited in the text). TATE'S (244) puts the land foot of Shanghai at either 33.5 cm (13.19 inches) or 13.2 inches. The Chinese used an unrelated foot of about 14 inches for most other purposes.

[26] TATE'S, 244.

[27] The individual values are given in SKINNER, 42–43.

[28] *General Laws of the Territory of Colorado,* 149.

[29] Hall, *Four Leagues of Pecos,* 85. In Texas, a law still on the books puts the vara at $33\frac{1}{3}$ inches, but this appears to have been adopted for convenience (making 108 varas equal to 100 yards) rather than for accuracy.

[30] UN (1955), 25.

[31] UN (1955), 27.

[32] MARTINI, 249. I'm assuming here that the klafter of Heidelberg was six feet long, as it was elsewhere in Baden-Württemberg.

[33] Bettess, "The Anglo-Saxon Foot," 50 (citing B. Hope-Taylor, *Yeavering: an Anglo-British centre of early Northumbria,* London, 1977, for the 281 mm foot found at Yeavering.)

[34] Kennelly, *Vestiges of Pre-metric Weights and Measures,* 161.

[35] Rounded off to 223 cm in UN (1955), 114.

[36] JOHNSON'S, 481.

[37] JOHNSON'S, 481.

[38] MARTINI, 115; UN (1955), 72.

[39] JOHNSON'S, 482.

[40] Mackay, *Further Excavations at Mohenjo-daro,* 1:404–06, and 2: plates CVI 30 and CXXV 1. See also SKINNER, 12.

[41] SKINNER, 40.

[42] Skinner's *Weights and Measures* provides an excellent overview of the survival of these ancient standards into the 19th century.

[43] CONNOR, 84.

[44] U.S. Department of Commerce, *Units and Systems of Weights and Measures,* LC 1035, 5–6.

Chapter 2
The troy pound

There has never been any doubt among historians that the troy pound is the oldest and most basic weight standard in the Customary tradition. In 1821, a British parliamentary commission found troy weight to be "the ancient weight of this kingdom, having... existed in the same state from the time of St. Edward the Confessor"[1]—that is, from before the Norman Conquest of 1066. In fact, as we shall see further on, the troy standard was already thousands of years old by that time.

Like the Northern foot, the troy pound represents the survival of an ancient tradition that suffered a collision in England with a standard from a completely different tradition, in this case the unrelated avoirdupois pound. As related in Chapter 5, this latecomer, the familiar everyday pound we still use in the U.S. today, is an old Roman city standard that came into medieval England from Italy as a unit for measuring wool, at that time England's chief export. The avoirdupois pound was originally used exclusively for wool, while troy in various forms was used for weighing almost everything else. Most importantly, the troy pound was for centuries the weight by which bread was sold, and by Elizabethan times it had become the basis by which the price of bread was regulated by the government.

Figure 2.1 shows the Exchequer standard troy pound of Elizabeth I. The standard, "of elegant form," and reputed, like the other Elizabethan weights of 1588, to be made from the melted down bronze ordnance of the Spanish Armada,[2] consists of a set of 12 nested "cup weights" ranging from 256 troy ounces (a bit over 20 pounds troy) down to $\frac{1}{8}$ troy ounce, plus a tiny solid extra $\frac{1}{8}$ ounce weight that fits in the center of the set.[3] There was no separate pound weight, the troy pound standard being formed by the combination of the 4 ounce weight and the 8 ounce weight shown here (the troy pound

Figure 2.1: Exchequer standard troy pound of Elizabeth I (1588)

is 12 troy ounces).[4] This set of 13 weights allows the verification, using a simple balance, of any troy weight up to 512 ounces ($42\frac{2}{3}$ troy pounds) in increments of $\frac{1}{8}$ ounce—over four thousand distinct standard values. The pure binary series exhibited by this set requires the fewest weights of any possible system that provides all the values in a given range[5] and demonstrates why all traditional systems of weight exhibit a strong binary component.

Troy weight remained the primary legal standard of weight in England until 1834, but by then it had long ceded practical application to avoirdupois for all except the most subtle measurements—of drugs, for example, and precious metals. Despite its limited range of uses, however, the troy standard would probably have remained the legal basis of weight in the customary system had not a catastrophe destroyed the physical standard itself.

Chapter 2. The troy pound

Figure 2.2: The troy pound of 1758

The destruction of the troy standard in Britain

Ironically, the loss of the troy standard was indirectly due to an early 19th century effort to overhaul the Customary System, which by then had become a farrago of conflicting standards from a number of different traditions. In 1824, the British Parliament decided to abandon several different standards of capacity in favor of a single "Imperial gallon" that would be based on an avoirdupois weight of water, a move clearly designed to ape the weight/volume relationship of the newly created French metric system (one liter of water weighs a kilogram, etc.). And it decided to base the whole system on just two physical standards: the Standard Yard of 1760 and the Parliamentary Troy Pound of 1758, both of which had been constructed after the most careful comparison with the best official standards then surviving. Even the avoirdupois pound was just a subsidiary weight defined in terms of the troy standard. The statute of 1824 enacted that

> ... the Standard Brass Weight of One Pound Troy Weight, made in the Year One thousand seven hundred and fifty eight, now in the Custody of the Clerk of the House of Commons, shall be and the same is hereby declared to be the original and genuine Standard Measure of Weight, and that such Brass Weight shall be and is hereby denominated the Imperial Standard Troy Pound, and shall be and the same is hereby declared to be the Unit or only Standard Measure of Weight, from which all other Weights shall be derived, computed and ascertained; and that One twelfth Part of the said Troy Pound shall be an Ounce; and that One twentieth Part of such Ounce shall be a Pennyweight; and that One twenty fourth Part of such Pennyweight shall be a Grain; so that Five thousand seven hundred and sixty such Grains shall be a Troy Pound, and that Seven thousand such Grains shall be and they are hereby declared to be a Pound Avoirdupois, and that One Sixteenth Part of the said Pound Avoirdupois shall be an Ounce Avoirdupois, and that One sixteenth Part of such Ounce shall be a Dram.

"Standard" meant then, as it did everywhere until quite recently, an actual, unique object with reference to which everything else was measured. Figure 2.2 shows what the troy pound of 1758 looked like.[6] By the statute quoted above, the avoirdupois pound was defined as something that weighs $\frac{7000}{5760}$ as much as this object.

By basing the Imperial measures on just two such reference objects—the yard of 1760 and the troy pound of 1758—Parliament greatly simplified the system. But in 1834, a fire consumed the Houses of Parliament, and with them the two official standards that had been kept by the Clerk of the House of Commons. The yard standard was irreparably damaged, and the troy pound of 1758 was never found.

Faced with the task of reconstructing the whole system of measurement from scratch, the government capitulated to popular usage and rebuilt the system of weight on the avoirdupois pound.[7] Just as in the the redefinition of the Northern foot in terms of a unit from a different tradition, however, this resulted in awkward and unnatural relationships between units deriving from one tradition and units deriving from the other.

Figure 2.3: J. M. W. Turner, *The Burning of the Houses of Parliament, 16 October 1834*. Oil on canvas, Cleveland Museum of Art (from Wikimedia)

In the troy tradition, the number of grains in an ounce is 480—a relationship that lends itself to a large number of useful whole-number subdivisions and echoes the similar ancient division of our mile into 4800 Northern feet. But the statute of 1824, the first to officially define avoirdupois in terms of troy, resulted in an avoirdupois ounce of $437\frac{1}{2}$ grains—a truly hideous definition that makes the division of a mile into 5280 feet look almost sensible by comparison.

The survival of troy weight in the United States

On the other side of the Atlantic, however, the fire effectively never happened, and the traditional system continued in full force. The Parliamentary Troy Pound of 1758 lived on in an exact copy that had been obtained by Albert Gallatin, United States minister at London, at the request of the director of the United States Mint at Philadelphia.

This copy was made in 1827—before the fire that destroyed the British standards—directly from the Standard Troy Pound of 1758 by the same persons who constructed the British standards authorized

Chapter 2. The troy pound

by the act of 1824, using the same equipment. The U.S. Troy pound was personally adjusted and verified by the physicist Henry Kater, a leading member of the commission of 1819 whose metrological work determined the length of the seconds pendulum and formed the basis for the official determination of the density of water. Thus, the copy (which still exists; see Figure 2.4) has the same metrological status as the ones that were created for implementing the troy-based 1824 reformation of the system.

The duplicate troy pound was enclosed in a casket and brought to the U.S. under seal by special messenger. There it was delivered, along with certificates from Kater and Gallatin attesting to its accuracy, to the Director of the Mint, in whose custody it remained, still sealed, until President John Quincy Adams could come from Washington to Philadelphia in order to verify its authenticity in person. In a special ceremony held October 12, 1827, the casket was unsealed, the documents accompanying it were verified, and the standard weight was officially declared to be identical to the Imperial standard troy pound of 1758.

Adams's finding of authenticity was transmitted to Congress through the Committee on the Mint, and on May 19, 1828, this troy pound was declared by an act of Congress to be "the standard troy pound of the Mint of the United States, conformably to which the coinage thereof shall be regulated." It remained the official coinage standard of the U.S. until 1911, when the troy pound of the Philadelphia mint was replaced by the troy pound of the U.S. Bureau of Standards.

Although the act of 1828 only specified its use for coinage, the troy pound of the Philadelphia Mint became the fundamental weight standard of the United States and the standard on which the avoirdupois pound was based, just as in England before the fire of 1834.[8] Brass weights such as this lose their accuracy over time due to oxidation, but when reweighed in 1910, the troy pound of the mint was found to be off by just $\frac{7}{1000}$ of a grain or one part in a million.[9]

Figure 2.4: The troy pound of the U.S. Mint, basis for all U.S. weights from 1828 to 1911

The troy system

The basis of both the troy and the avoirdupois systems of weight is the troy grain, which is the grain in a 5-grain aspirin tablet[10] or a 170-grain rifle bullet. By current law in the U.S., the troy grain weighs 64.79891 milligrams, exactly. Thus, the grain and its multiples are now derived units of the metric system. But the number chosen really does preserve our traditional standard to seven places of accuracy and can reasonably be said to carry this ancient cultural artifact forward intact.

The troy system of weight, still legal in the U.S. today, is as follows:

1 pennyweight (dwt) = 24 grains

1 troy ounce (oz) = 20 dwt or 480 grains

1 troy pound (lb) = 12 ounces = 5760 grains

The troy ounce and pound are substantially different from the common avoirdupois ounces and pounds that we use for buying groceries. The troy ounce of 480 grains is somewhat heavier than the avoirdupois ounce of $437\frac{1}{2}$ grains, but the troy pound is quite a bit lighter than the avoirdupois pound because the troy pound contains only 12 troy ounces or 5760 grains, compared with 16 avoirdupois ounces or 7000 grains for the avoirdupois pound. Dividing a pound into 12 ounces is a Roman tradition, whereas dividing a pound into 16 ounces is a Germanic tradition.

Instead of the pennyweight ($\frac{1}{20}$ ounce or 24 grains) used by jewelers and goldsmiths, druggists in the old days commonly used the apothecaries' dram ($\frac{1}{8}$ oz or 60 grains) and its third, the scruple ($\frac{1}{24}$ oz or 20 grains). So for them, the system was

1 apothecaries' scruple = 20 grains

1 apothecaries' dram = 3 scruples or 60 grains

1 (troy) ounce = 8 drams or 480 grains

The division of an ounce into scruples and drams is very old, the scruple being our version of the Latin *scripulum* and the dram (or drachm, as it's spelled in the UK) being our version of the Greek *drachma*.

There was also a troy unit of weight called the shilling, which survived until recently in England as the name of a coin but was also for many years (including a couple of centuries in the colonial U.S.) the name of an actual unit of weight, equal to $\frac{1}{20}$ of a troy pound or 288 grains.

Modern uses of troy weight

The troy standard remains the preferred measure in the United States for weighing silver, gold, gunpowder, and bullets. Over the last half century, most of its other uses have passed into history, but these uses are enough to maintain its existence on most electronic scales.

The troy unit most people are familiar with is the troy ounce, which is still the primary unit of measure for precious metals. This is the ounce in price quotations for gold, silver, and platinum. The troy ounce also remains our official measure for gold and silver coins in the U.S., which are minted in denominations of one troy ounce and its fractions.

Some foreign coins are also minted in units of troy weight, obscured, however, by a bureaucratic insistence on expressing these weights in apparently meaningless metric equivalents. In a flier advertising two Mexican proof coins, for example, it is stated that the 100-peso silver coin weighs 33.625 grams of .925 fine silver and the 250-peso weighs 8.640 grams of .900 fine gold; it takes some work with a calculator to discover that the former simply contains one troy ounce of pure silver and the latter, $\frac{1}{4}$ troy ounce of pure gold. The troy pound of 12 troy ounces is still occasionally used in the U.S. as a unit of measure for certain large coins and commemorative medals sold to collectors.

The troy grain notionally represents a grain of barley,[11] which is roughly the same weight. This unit is perfectly suited to the weighing of bullets and gunpowder, since the quantities typically used can be expressed without loss of precision in reasonably small integer numbers of grains. The old carat for weighing precious stones was three troy grains,[12] but that has now been entirely replaced by metric units.

From the legal definition of the troy grain as exactly 0.06479891 grams, you can easily convert between customary weights and metric weights. If you have a calculator that can store constants, enter 0.06479891 as one of them (you could call it C, for example). To convert grams to grains, divide by C; to then convert this figure to pounds, divide by 5760 for troy pounds or 7000 for avoirdupois pounds. To convert grains to grams, multiply by C; to then convert this figure to kilograms, divide by 1000. All of the resulting figures will be exact to as many places of accuracy as your calculator will support.

While the troy pennyweight of 24 grains was not the actual weight of an English silver penny (which, as recounted in Chapter 4, was a little lighter), the coin did give us the d in the abbreviation dwt; d stands for *denarius*, an old Roman coin of similar size. Still used to some extent by jewelers and goldsmiths, the pennyweight is a handy unit for small, valuable objects such as gold rings.[13] It is exemplified in a standardized object you may still encounter occasionally—the U.S. bronze cent, known formally as the Lincoln cent. From its introduction in 1909 to its replacement by cheaper alloys based on zinc in 1982, the U.S. bronze cent weighed exactly 48 grains or two troy pennyweights. Ten bronze Lincoln cents weighed one troy ounce, and 120 of them weighed one troy pound. The 48-grain unit represented by the Lincoln cent has quite a history, as you will see.

Chapter 2. The troy pound

The ancient history of troy weight

Limited as its use today has become, troy weight is still deserving of respect if not downright awe; the troy standard of weight is probably the oldest continuously maintained human cultural artifact in existance.

As indicated by names such as pound (pondus), ounce (uncia), and scruple (scripulum), the system of troy weight divisions that has come down to us in the Customary system is basically Roman, and dates in England from the Roman occupation in the early part of the Christian era. The standard itself, however, is almost four thousand years older. The troy standard appears to have originated in Egypt, whence it was spread thoughout the ancient world by the Greeks. Remains of it can be found in the premetric 19th century standards formerly used over large areas of Europe, Africa, and Asia, as will be documented further on in this chapter.

To follow this history, however, it's necessary to understand that what I am calling the troy standard is actually a web of interlocking weight units bearing simple whole-number relationships to each other. Depending on the culture in which the system was used, some of these weights were given their own names and considered the units of the local system, while the others were just considered simple multiples and submultiples of the named units. Thus in our version of the troy system, the weight of $\frac{1}{20}$ troy pound (288 grains) was for centuries given its own name—the shilling. This name went out of common use in the U.S. around 200 years ago, but of course the weight, as the twentieth of the pound, did not; we just stopped thinking of it as a unit of the system. Similarly, when the British abandoned the troy pound as a named unit in the 19th century while keeping the troy ounce, the quantity of 12 troy ounces did not cease to exist; it simply stopped being used in Great Britain as a named unit. So in the account that follows, I will often be pointing to equivalences between named units in one place that correspond to simple multiples or submultiples of named units in some other place to show the spread of the troy standard across a large part of the ancient world. I have included diagrams that I hope will help illustrate these equivalences.

It will also be useful here to take a quick look at a few very general features of ancient systems of weight.[14]

The notional foundation of all ancient systems of weight is a seed of some kind. In the English troy tradition we've inherited, that seed is a grain of barley. In reality, all known ancient weight standards were actually defined within a set of relationships that specified both weights and measures of capacity based on a certain bulk weight of grain. In the most ancient systems, the basis was bulk weights of barley; our system, as you will see, was based on bulk weights of wheat.

The next major unit up from the grain in ancient systems of weight is generically known as a shekel. Notionally, a shekel is one handful of some kind of grain; in actuality, by the late bronze age all known ancient shekels conformed to a particular standard of weight set up by the ruling royalty or priesthood. Following the invention of coinage around 700 B.C., the shekel acquired additional meaning as the weight of a coin called a didrachma, which was twice the weight of the most common coin, the drachma. There were also a number of smaller weight units considered subdivisions of the shekel, the most important being the $\frac{1}{2}$ shekel and the $\frac{1}{3}$ shekel. Following the Romans, double or quadruple shekels came to be called by some name derived from the Latin uncia (meaning "unit")—in English, ounces.[15]

Virtually all ancient systems of weight had a primary unit about the size of a pound or half kilo. The bronze-age units of this size are known to archaeologists as minas. Minas were 25, 30, 40, 50, or 60 shekels of a particular weight system; for example, the Sumerian mina was 60 shekels of about 129.6 troy grains or 8.4 grams and weighed 503 to 505 grams, about 11 percent larger than our U.S. pound.

Finally, all the ancient weight systems had a unit of 60 minas generically called (by archaeologists) a talent. This is the same talent referred to in the Bible, and it constituted the chief weight unit for cargoes. A talent is about the size of a load (somewhere around 60 pounds) that a person could carry for distances of up to about 400 yards, beyond which a pack animal would have to be used.[16]

In general, ancient systems of weight were built conceptually from the grain upward, but historically, it's more probable that the load (talent) was the first unit of weight to be standardized. As the Assyriologist Thureau-Dangin observed,

> For a long time one had to count exclusively by "loads" (in Sumerian *gun*, in Accadian *biltu*). Apparently, the need for a smaller and a more precise unit of weight made itself felt only at the time when the metals came into general use and became an article of trade. Only at that time, one was compelled to construct the first balances, to use the first weights and to establish a fixed ratio between the old "load" and the new, purely conventional, unit, which one called by the name *mana*, "a mina." The secondary character of the mina is clearly indicated by the fact that the Sumerian script has no ideogram to represent this unit. One wrote phonetically *ma-na*. If one assumed a sexagesimal ratio between the "load" (that is the talent) and the mina, it was doubtless due to the fact that the number 60 was already the base of numeration.[17]

A history of weight standardization working down from the talent would also explain why all the Middle Eastern systems of weight divided the talent into 60 minas but then went in several different directions when dividing the mina into smaller units.

Troy weight was the most important but by no means the only ancient weight standard; others included the Sumerian standard mentioned above and the Greek standard of Athens, which we'll take a look at in Chapter 4 in connection with our units for measuring dry substances. More often than not, these standards were used in parallel as merchants and moneychangers brokered trade goods and money between different locations; consequently, finds of ancient weights almost always contain a confusing mixture of different units based on two or three different standards. Presenting a coherent account of the troy standard requires a particular focus on the data, and I hope the reader will understand that we are looking at selected guideposts in a very complicated landscape.

The basis of the troy weight standard is Egyptian, so that's where we'll start.

Egyptian origins of the troy standard

The earliest known set of weights conforming to any standard is a collection of six alabaster cones found by Flinders Petrie in First Dynasty Egyptian graves at Tarkhan, dating from around 3800 B.C. These six objects come from four graves, four of them found in pairs and two singly. The weights and interpretations are reported as follows:[18]

Grave No.	Weight (grains)	Divisor	Unit
1548	845.3	18	47.0
717	478.2	10	47.8
717	144.8	3	48.3
728	872.6	18	48.5
728	985.0	20	49.2
1892	980.0	20	49.0

Even in this earliest example we see the forerunner of the troy ounce (480 grains) in the weight of 478.2 grains and its rough double in the weights of 980.0 and 985.0 grains.

The ur-unit of 48 grains shown in this most ancient of weight sets manifested itself in two closely related Egyptian traditions. One, the weight system of ordinary commerce, was based on a shekel called the qedet of about 144 troy grains,[19] i.e., 3×48 (note the weight of 144.8 in the earliest set above); the other, always associated specifically with the weighing of gold, was based on a shekel called the beqa of about 192 troy grains, i.e., 4×48.[20] Weights of about 5760 grains—that is, 40 qedets or 30 beqas—are fairly common in the ancient data;[21] this is, of course, our troy pound.

The ancient Egyptians didn't use the troy pound as a named unit, though. They liked their systems decimal, so the weight system seen in Egyptian manuscripts is 10 qedets = 1 deben (1440 grains) and 10 debens = 1 sep (14,400 grains). A very widely used but unnamed unit of 50 shekels (half a sep or 7200 grains) was their version of a pound.

Being the earliest examples known, the Predynastic weights from Tarkhan listed above are not terribly precise. That the technology got much better over time is clear from the 1961 excavation of a bronze-age shipwreck off Cape Gelidonya on the coast of modern Turkey, which turned up a wonder-

Chapter 2. The troy pound

fully preserved set of stone weights belonging to the ship's merchant captain. Dating to the 15th century B.C., the weights correspond accurately to several well-attested ancient standards.

Eleven of the 52 weights are based on the Egyptian qedet of 144 grains. Their values show the accuracy with which the weights were maintained by a real captain of a real ship carrying copper ingots from Cyprus. Here they are, with the weights as reported in grams on the left and their calculated troy equivalents on the right:

Weight (grams)	Weight (grains)	Number of qedets	Weight of qedet (grains)
470	7253	50	145.1
468	7222	50	144.4
279.5	4313	30	143.8
233	3596	25	143.8
188	2901	20	145.1
185.5	2863	20	143.1
92	1438	10	143.8
56	864	6	144.0
55.5	856	6	142.7
28	432	3	144.0
9.3	143.5	1	143.5

Only the last of these is marked, showing that its owner considered it the unit of the system.[22] Taken together, these weights give an average value for the Egyptian qedet of 9.328 grams or 143.95 troy grains. Forty of these qedets would weigh 5758 grains, a value that differs from the modern troy pound of 5760 grains by less than one part in a thousand. If we normalize this set to a qedet value of 144 troy grains exactly, the weights of this standard carried on the ship would have values of 144, 432, 864, 1440 (the deben), 2880, 3600, 4320, and 7200 (half sep). We will see these weights occur repeatedly as we continue our history.

The study of weights as old as these is complicated by slight variations between the version of a particular standard in each location and much larger variations caused by damage or corrosion, which of course directly affects the weight we get when we weigh them today. Thus, a collection like this, composed of relatively well preserved stone weights taken from a single location, is of particular significance.

Another bronze-age shipwreck off the same dangerous coast of southern Turkey, this one near the town of Uluburun and dating from 1305 B.C. or shortly thereafter, was found in underwater excavations between 1984 and 1994 to contain 149 objects classified as balance weights. The metal weights were eliminated from consideration due to corrosion; each of the stone weights was then painstakingly reconstructed by finding the volume of plasticine needed to restore chips from its surface and calculating the missing weight based on the density of that particular variety of stone. This technique, developed by Flinders Petrie a century earlier, is the only one that yields accurate results for weights that are not in a perfect state of preservation.

As with the weights from the Cape Gelidonya shipwreck, the ones found at Uluburun correspond to several different bronze age standards, but weights of the qedet standard predominate. These can be divided into two subclasses based on their shape. The first subclass consists of weights of a "sphendonoid" shape showing an average unit weight of 9.35 gm (144.3 grains) and a weighted average value (favoring the heavier weights) of 9.33 gm (144.0 grains); the second, consisting of weights having a domed shape, show an average unit weight of 9.34 gm (144.1 grains) and a weighted average of 9.32 gm (143.8 grains).[23] The denominations of the qedet unit found in the wreck are reported as $\frac{1}{2}$, $\frac{2}{3}$, 1, 2, 3, 5, 10, 20, 30, 50 and 100; if we take the unit itself as 144 grains (9.331 grams), then these weights would correspond to 72, 96, 144, 288, 432, 720, 1440, 2880, 4320, 7200, and 14,400 grains.

The shipwreck data is some of the best we have relating to ancient standards of weight, and combining it with documentary evidence gives us a clear picture of the dominant middle Eastern weight standards of the late bronze age. The mathematical relationship between three of the most important weight standards of the era—the Egyptian, Hittite, and Palestinian/Hebrew shekels—can be deduced from treaties and correspondence in the period of about 1350–1230 B.C. among the rulers of Egypt, Ugarit, the Hittite empire, and the Canaanite city of Gezer. Analysis centers on a unit known as the "mina of Carchemish," referred to in a number of

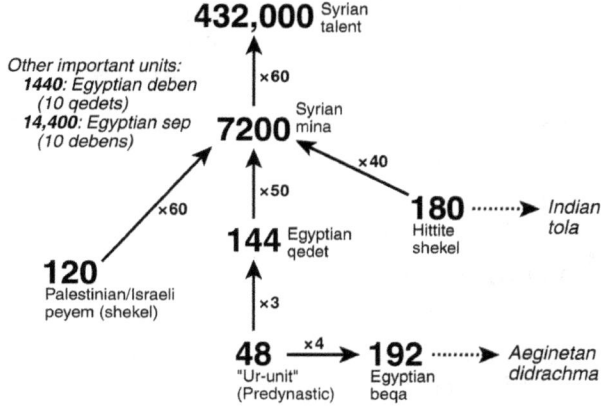

Figure 2.5: The dominant late bronze-age weight system as shown in Carchemish (all weights in troy grains)

cuneiform documents discovered beginning in the late 19th century.

The Hittite city of Carchemish was a crossroads of trade on the river Euphrates[24] and therefore the scene of everyday conversions between units of the major national weight systems. It can be established from various pieces of correspondence that the mina of Carchemish was alternatively divided into 40 Hittite shekels, 50 Egyptian qedets, or 60 Palestinian/Hebrew shekels or peyems,[25] and that 60 of these "Syrian minas" (i.e., 2400 Hittite shekels, 3000 Egyptian qedets, or 3600 peyems) constituted the Syrian (Egyptian/Hittite) talent.[26]

As the single most important and probably the oldest bronze-age standard of weight, the Egyptian qedet is the linchpin of the whole complex, and the shipwreck data nails down its value and by implication the values of the other two standards to a precision of better than one part in a thousand. Figure 2.5 provides a graphic representation of the system's major units and their interrelationships. It's clear from the equivalence with 50 qedets that the Syrian mina or "mina of Carchemish" was the half sep, with a weight (based on the shipwreck data) of 7200 troy grains.

Troy weight in the Iron Age

While clearly related to our troy standard (and many later units, as we shall see), the named units of the Syrian (Egyptian/Hittite) system do not yet include the troy units that have come down to us. It is in the first Greek system of weights and measures, established in the 7th century B.C. by Pheidon, King of Argos,[27] that we explicitly see the troy pound itself and its primary divisions. Table 2.1 shows the Argive weight system of Pheidon according to a recent authoritative summary of Greek weight standards (I've added the three columns on the right to show the troy equivalents).[28]

The archaic Greek half stater of 374.22 grams is none other than our troy pound, about 0.3 percent heavier by this estimate (5775.1 troy grains instead of our 5760).[29] If we adjust all the units by this amount, we get the perfect series of troy equivalents listed in the last column. Notice that the $\frac{1}{8}$ stater unit corresponding to 3 troy ounces is the common Egyptian deben (10×144).

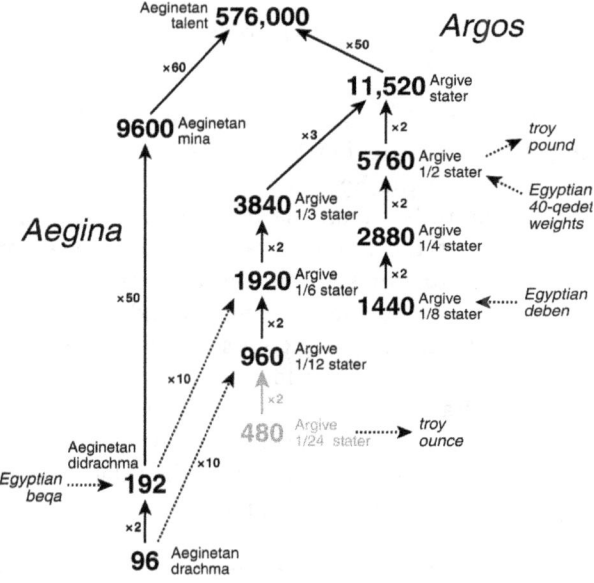

Figure 2.6: Early Greek weight systems

The inheritance of Greek standards from the Egyptian system is one aspect of the often observed transfer of Egyptian culture to early Greece, and in fact Argos was not the only Greek city to form a sys-

Chapter 2. The troy pound

Unit	Weight (grams)	Weight (grains)	Troy cognate units (0.3% difference)	Troy cognates in grains
Double stater	1496.88	23,100.4	48 troy ounces (4 troy pounds)	23,040
Stater	748.44	11,550.2	24 troy ounces (2 troy pounds)	11,520
$\frac{1}{2}$ stater	374.22	5775.1	12 troy ounces (1 troy pound)	5760
$\frac{1}{3}$ stater	249.48	3850.1	8 troy ounces ($\frac{2}{3}$ troy pound)	3840
$\frac{1}{4}$ stater	187.11	2887.5	6 troy ounces ($\frac{1}{2}$ troy pound)	2880
$\frac{1}{6}$ stater	124.74	1925.0	4 troy ounces ($\frac{1}{3}$ troy pound)	1920
$\frac{1}{8}$ stater	93.56	1443.8	3 troy ounces ($\frac{1}{4}$ troy pound)	1440
$\frac{1}{12}$ stater	62.37	962.5	2 troy ounces ($\frac{1}{6}$ troy pound)	960

Table 2.1: The oldest Greek weight system (Pheidon of Argos, 7th century B.C.)

tem based on Egyptian weight standards. Another Greek city-state, Aegina, formed its own weight system based on the Egyptian gold shekel (beqa) of 192 grains rather than the qedet of 144 grains.[30] The basis of the system was the half shekel or drachma of 96 grains;[31] this is just two "ur-units" of 48 grains. The Aeginetan drachma and didrachma were among the primary coin standards of the ancient Mediterranean world. The Aeginetan mina was 50 didrachmas (9600 troy grains), and the talent was, as usual, 60 minas, weighing in this case 576,000 troy grains. A diagram of the relationships between the systems of Aegina and Argos is given in Figure 2.6.

There is a long history of unsuccessful attempts to resolve the etymology of the word "troy" in "troy weight," ranging from a proposed descent from the name of the Medieval French city of Troyes to a survival of the word Troy used as a synonym for London;[32] but the perfect alignment of troy weight with the earliest Greek systems makes it at least possible that the name Troy actually refers to the ancient city that Homer wrote about.

The little evidence we have regarding the weight system of Troy supports this: among 39 weights from Troy reported by Schliemann, 10 belong to the qedet standard and show an average value for the qedet of 144.5 grains.[33] In its later incarnation as Ilium, Troy remained an important city well into the Byzantine era, so the influence, if there was one, had a couple of millennia to make itself felt.

The source of the troy standard is easier to identify than the origin of the name. Greeks of the classical era engaged in trade with Britain, a primary source of tin,[34] and it seems very likely that the troy standard came into Britain as a result of this trade.

A small but significant collection of Etruscan weights shows the penetration of troy into the Italian peninsula (though troy was not used by the later Romans). Of 25 Etruscan weights found at a site near modern Bologna, 17 (15 of them marked) are based on a unit of about 576 grains, and of the 15 marked examples, eight are either 1, 10, or 100 of this unit. The average value of the unit evidenced by the 17 weights belonging to this standard is 576 ±2 grains; if one outlier is removed, the remaining 16 average 578.7 ±1.7 grains.[35]

Spread of the troy standard

The Greek standard was spread by trade to the south, in particular to the great medieval Arabic civilizations, and came to be preserved in many traditional European, African, and Asian systems of weight well into the 19th century. Shaped by its collision with even older indigenous systems, often based on different notions of counting and weighing, the standard was expressed in a number of different units that can be seen as equivalent by comparing corresponding multiples and submultiples of the base units in each place.

Survivals of the troy standard into the modern era can be divided, somewhat artificially, into two families: one in which the units are multiples or submultiples of a base unit weighing 48 (or 480) grains and one in which the units are multiples or submultiples of a base unit of 576 (or 57.6) grains.

The 576/57.6 branch can be seen as representing a duodecimal multiple of the ur-unit ($576 = 12 \times 48$), or a binary multiple of the Egyptian qedet ($576 = 4 \times 144$), or, perhaps most usefully, as a decimal division of the unit we call the troy pound ($576 = 5760 \times \frac{1}{10}$) and the older Aeginetan talent ($576 = 576,000 \times \frac{1}{1000}$).

The troy survivals based on a unit of 576 (or 57.6) grains are most clearly seen in the apothecaries' systems of premetric Germany, Denmark, and Russia. In the pharmaceutical system of Nuremburg, used in Denmark and throughout the German empire, the 19th century value of the "drachme" was 3.7269 grams or 57.514 grains—a hundredth of the troy pound at about 0.15 percent below the English standard.[36] In Russia, on the other hand, the medicinal dram was 57.602 troy grains, which is the hundredth of a troy pound almost exactly.[37]

While it pops up occasionally in our story, this decimal family (with the exception of the troy pound itself) is seen much less often than multiples of 48 and 480. The latter family is most clearly exemplified by the Arabic ukiah (ounce) for gold of about 480 grains and its tenth, the gold bullion dirhem (dram) of about 48 grains. These units are widely attested in medieval Arabic weights, and the troy ounce in its various later European incarnations is almost certainly a descendent of the Arabic gold ukiah.[38]

Both of these families are attested in the Saxon weights that led up to the troy system as we have inherited it.

Troy weight in Saxon England

The troy system was first defined by name relatively late in English law by the statute 12 Henry 7 c. 5 (1496), which declared that every troy pound

> conteyne xij unces of troy weight, and every unce conteyn xx sterlinges [pennyweights], and every sterling be of the weight of xxxij cornes of whete that grewe in the myddes of the Eare of the whete according to the old Lawes of this Land.[39]

In this passage, the phrase "according to the old laws of this land" refers to the time before the Norman Conquest, more than four centuries earlier, and it shows that the statute of 1496 was merely rehearsing a system that had been in use from time immemorial. The reference to a basis in grains of wheat is ritual; in fact, as Queen Victoria's Warden of the Standards stated in 1873, "there is no record whatever of the actual construction of any standard unit of troy or avoirdupois weight based upon the weight of 32 grains of corn."[40]

The pre-conquest (Saxon) history of troy weight in the British Isles was revealed early in the 20th century with the discovery of several dozen ancient weights of the troy standard at various locations in England and Ireland. That most of these weights were found in graves demonstrates how important they were to their owners. Even at this relatively late stage of its history, the troy standard still exhibited both the 48-grain and 576-grain modes of division.

The most abundant of the British and Irish weights represent multiples or subdivisions of the 48-grain unit. Thirty weights from 6th and 7th century graves of Saxon moneychangers and 9th century Viking ship burials in Ireland listed in a 1923 survey[41] give an average of 48.43 grains for the base unit.[42] The fact that two of these weights are from Dublin and two from an Irish/Viking ship burial suggest that the unit was already in use in the Celtic tradition before the Saxons arrived in Britain. One 7th century Saxon moneychanger's set contains weights of 8, 7, 6, 5, 3, 3, and 2 units of a standard with an average value of 48.79 grains; another set (see Figure 2.7) has weights of 6, 5, 4, 3, 3, 2, 2, 1, $\frac{1}{3}$, and $\frac{1}{6}$ of a unit averaging 47.88 grains.[43]

A smaller but still significant number of English and Irish weights are multiples or submultiples of the 576-grain and 57.6-grain units, and again there is evidence—a weight of 574 grains bearing a fleur-de-lis design on top, found at the site of a Roman camp at Melandra, Derbyshire—that this unit was extant in Britain before the arrival of the Saxons.

The finest example of the decimal subdivision of the troy pound in Anglo-Saxon times is a perfectly preserved bronze weight found near Grove Ferry in Kent.[44] It was weighed in 1922 and found to be "exactly 576 troy grains" or $\frac{1}{10}$ of a troy pound.[45] In Silchester, an Anglo-Saxon weight of 5761.7 grains—almost exactly the troy pound—was accompanied by two weights showing a decimal division of the pound, two tenths in one case and eight tenths in the other.[46]

Six of the Saxon weights prove the existence in England of a unit of 144 grains, the ancient Egyptian qedet itself. Two of these are especially significant because they are among the very small minority of all ancient weights to bear markings indicating their value. One, from a grave at Gilton, Kent, weighs 144 grains and is marked 3; the other, from a grave at Ozingell, Kent, weights 145 grains and is marked both 3 and 1, showing that the qedet itself still existed as a unit and that the 48-grain unit was considered its third.[47]

The late English and colonial American shilling of 288 troy grains was thus two Egyptian qedets, and the evidence of the weights dating from late antiquity in England and Ireland demonstrates that this was no mere coincidence but rather the survival of a standard whose continuous use spans almost six thousand years.

The haverdepois pound

The troy pound of 12 troy ounces (5760 troy grains) was not the only troy-based weight used in medieval England. There was also a unit of 16 troy ounces (that is, 7680 troy grains) known, confusingly, as the "haverdepois" or "Habur de Peyse" or "Haburdy Poyse" pound.

Thus, an anonymous 15th century tract titled *The Noumbre of Weyghtes* says of troy weight that it is used to buy and sell

> Gold, Sylver, Perlys and odyr precius stonys, and iuwells and beeds schold be sold by this weyght

whereas "Habur de Peyse" weight

> conteynyth xvi unces of Troy weyght, and by this weyght be all maner of merchaundyse bought and sold, as tynne, lede, iron, coper, style, waxe, wode, alem, mader, spycery, corsys [ribbons], laces, sylks, threde, flex, hempe, ropys, talowe, and such odyr as be usyd to be sold by weyght....[48]

Note the absence of wool from these lists; this will be important later.

Two things about the haverdepois pound have caused confusion among historians. The first is the name itself. The word *haverdepois* is obviously a variation on the same phrase that gave us "avoirdupois,"[49] and some sources even now conflate the haverdepois pound of 7680 grains with later references to the much smaller avoirdupois pound of about 7000 grains. This is understandable but incorrect; as stated earlier, and as I'll discuss in more detail in Chapter 5, our avoirdupois pound didn't yet have that name, and at this time it was used exclusively for weighing wool. The larger haverdepois pound is a distinct unit with its own history, as proven not just from documentary evidence but also by the existence of the same unit in Scotland and some of the northern European countries closest to England.

The Scotch version of this unit, known as the Scotch troy pound, was put by popular tradition at $17\frac{1}{2}$ of our modern avoirdupois ounces;[50] this is 7656.25 troy grains, making the Scotch troy ounce

Chapter 2. The troy pound

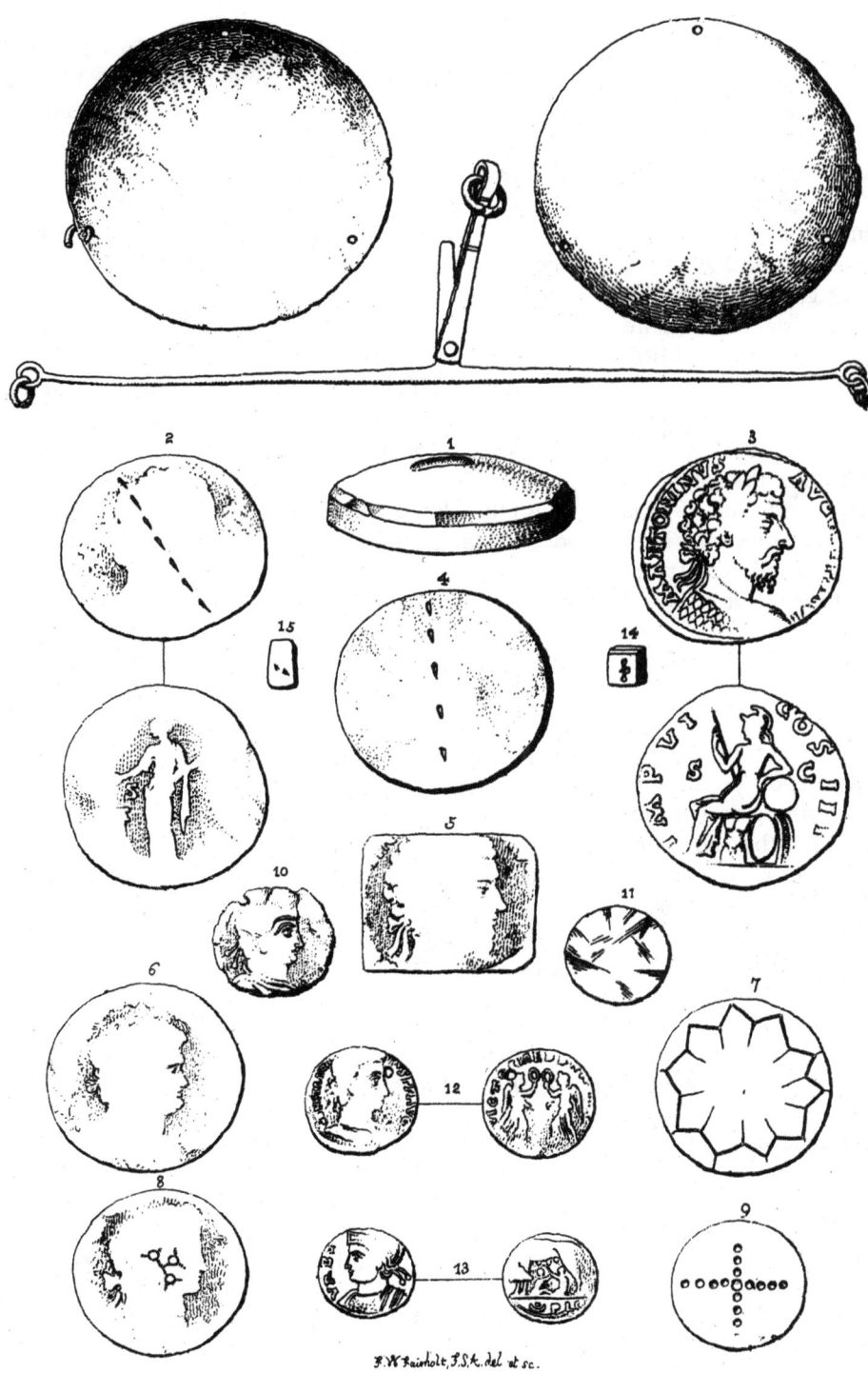

Figure 2.7: A 7th century Anglo-Saxon moneychanger's balance and weight set (scale pans at top). Most of the weights were made by grinding down coins of the period to the proper size

Chapter 2. The troy pound

478.516 English troy grains or about 0.31 percent less than the English standard.

Lying a roughly equal amount on the other side of the English troy standard was the premetric pund of Denmark, a unit whose weight of 499.309 grams[51] or 7705.52 troy grains was divided into 16 onces of 481.595 grains, about 0.33 percent above the English value. At 498.212 gm[52] the 19th century pund of Norway was even closer—7688.59 grains, divided into 16 unse of 480.537 grains, a bare 0.1 percent above troy. Another haverdepois cognate survived into the 19th century as the pfund of Bremen (Germany) at 498.5 grams[53] or 7693 grains, with an unze of 480.8 grains.

It's clear, therefore, that the haverdepois pound properly speaking actually existed, that it actually contained 16 troy ounces, and that it was a unit originally entirely distinct from the later avoirdupois pound. How its name came to be transferred to an unrelated standard for weighing wool will be explored further in Chapter 5.

While the existence of the old haverdepois pound as a unit distinct from the later avoirdupois pound is beyond question, it's not entirely clear what this weight was actually used for. According to the passage quoted above, the haverdepois pound was used for weighing just about everything except precious metals, precious stones, and (by omission) wool; this includes "spycery." Most dictionaries equate *spycery* with "spices," but it was also used in Old French to mean "groceries."[54]

In a comprehensive review of all English weights and measures of the time given in the *Coventry Leet Book* for 1474, it is stated that

> xxxij[ti] graynes of whete take out of the myddes of the Ere makith a sterling peny & xx[ti] sterling makith a Ounce of haburdepeyse; and xvj Ouncez makith a li., to be devyded from the most part in-to the lest part to by & sell Spycers by and all her Chaffer þat [all their trade that] perteyneth vnto weyght...[55]

Courts-leet were given jurisdiction over weights and measures by Edward II in a statute of uncertain date titled *Visus Frankiplegii* (View of Frankpledge), traditionally cited as 18 Edward 2 (1325). So this passage, which summed up the entire system of the time for weights of a pound or less, was intended to be official and categorical. The statement that 20 pennyweights (a unit that has never been associated with modern avoirdupois weight) make an "Ounce of haburdepeyse" and 16 of these ounces "makith a li." (pound) shows that this passage is referring to the haverdepois pound of 16 troy ounces.

The obvious interpretation of the word *spycers* is "spices" or perhaps "sellers of spices," but this is inconsistent with the fact that this "haburdepeyse" weight is the only pound recognized by the court for *any* purpose. Thus, the phrase "Spycers ... and all her Chaffer þat perteyneth vnto weyght" has to be interpreted as describing trade goods in general, and *spycers* therefore can only mean something like "groceries" (or perhaps, given the strained syntax, "grocers").

A table dated 1496 that used to hang on the wall of the Star Chambre at Winchester states that

> twelve uncs maketh a pounde of Troye weight for silver, golde, breade, and measure [i.e., capacity measures, as will be explained in Chapter 3].... XVI uncs of Troie maketh the Haberty poie a pound for to buy spice by....[56]

According to H. W. Chisolm, speaking *ex cathedra* as Queen Victoria's Warden of the Standards,

> This word is undoubtedly written "spice" in the MS., but it is evident from the context, as well as from the statute 24 Hen. 8. c. 3 ... that the word *spice,* like the German *speiss,* is intended to signify "meat."

The statute 24 Henry 8 c. 3 (1532) referred to by Chisolm is titled *An Acte for Fleshe to be sold by weight,* and it requires that

> every persone, whiche shall sell by hym self, or any other, the Carcases of Beoffes Porke Mutton or Veale or any parte or parcell therof ... shall sell the same by laufull weighte called Haberdepayes & no otherwise.

The appearance of this statute within living memory of the Winchester memorandum of 1496 makes the interpretation of *spice* in the earlier document as "meat" an attractive one; but German *Speiss* or *Speise* actually means food in general, not just flesh, and more to the point, there is no recorded use of *spice* meaning "meat" in Middle English (though the English word *meat* back then did indeed mean food in general). It is conceivable that the word *spice* in the Winchester memorandum is actually a variant spelling of Anglo-Saxon *spic* "bacon" (*spichus* "larder, meat-house," Old High German *spëcch*, Middle High German *spëc(k)*, "bacon, lard, fat," German *speck*, "bacon"); it survived into the 20th century as *spick* in the dialect of Shetland and the Orkney Islands with the restricted sense of "animal fat" (following the Old Norse meaning of the word)[57] and as *speck* in the U.S. with the meaning of "fat meat" or "blubber."[58] But the two previous citations make it more likely that *spice* and *spycery* derive from some folk blending of the Old French *espicier* "grocer" or *espicerie* "groceries" and the Old High German *spîsa* "food," which come from different Latin roots. By the time of Elizabeth I, this meaning had apparently been lost; her proclamation of June 1583 mandated that "Troy Weight... ought to be used only for the weighing of Gold, Silver, Bread, and Electuaries" and "Avoirdepois Weight" (meaning by then our modern avoirdupois pound) "is to be used for the weighing of Spices, and all other Things vendable by Weights."[59]

The pound of Cologne

The German and Scandinavian tradition represented in England by haverdepois weight divided its pound unit into 16 ounces corresponding to the troy ounce.[60] Multiplying the troy ounce by 15 instead of 16 yielded an even more important German pound of about 7200 grains. This other pound can be considered a survival of the Syrian mina of 50 Egyptian qedets, just as the troy pound is a survival of 40 qedets. As noted earlier, in places like Carchemish that used the system outside of Egypt, this was a named unit, the "mina of Carchemish"; and even in places where it was not a named unit, it existed as half an Egyptian sep in the form of real weights (as in the two shipwrecks described above) for two or three thousand years before the Germanic peoples encountered it via trade with the Mediterranean. The Germans divided this mina not into 15 troy cognate ounces, however, but into 16

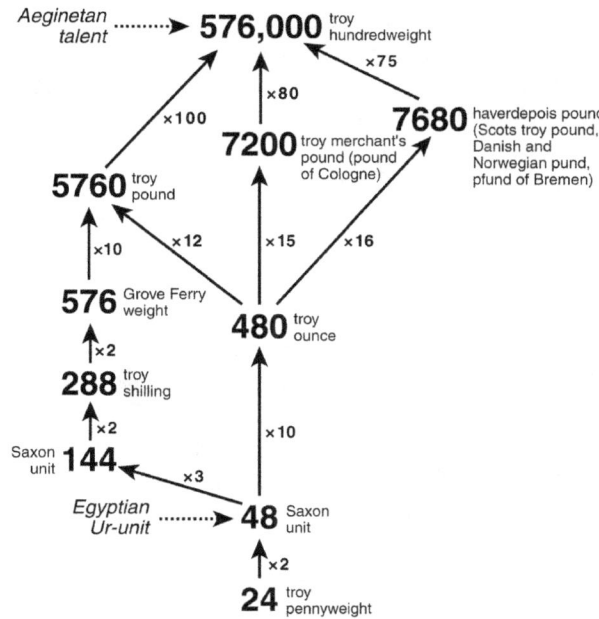

Figure 2.8: Troy weight in Saxon and medieval England

smaller units, and that is the form in which it survived through medieval times and into the 19th century German states.

Historically, the German survival of the half sep or Syrian mina is known as the pound of Cologne. It served as the primary unit of weight of the Hanseatic League, an association of German states bordering the North Sea that dominated commerce in northern Europe, including England, from the 13th through 16th centuries. The half pound or mark of Cologne was the chief northern European coinage weight.

So important did this Hanseatic standard become that it was finally made the subject of an international agreement by the Treaty of 1560. Three official standard weights made pursuant to this treaty are still in existence and give an average value for the pound of Cologne of 7219.31 grains, about 0.27 percent greater than the troy equivalent (15×480).[61] The premetric pfund of Prussia was 7218.26 troy grains,[62] also a bit higher than the troy equivalent, but 16 19th-century northern German city standards cited in a 1856 study average 7201.6 grains.[63]

Chapter 2. The troy pound

While all of these pounds are obviously 15 of the same unit we call the troy ounce, their division into 16 ounces according to the common Germanic custom resulted in an ounce of about 450 troy grains. This unit has its own history in England and will come back into our story in Chapter 4.

The relationships among all the major units of the Saxon and medieval troy weight systems are summarized in Figure 2.8.

Troy weight in the Arabic world

In Chapter 1, we saw that the Northern foot survived in Old World premetric length standards ranging from England in the west to China in the east. An examination of premetric measures of weight surviving into the 19th and 20th centuries reveals an equally wide range for the ancient standard we know as troy. Establishing this ubiquity is important to historical metrology because it places troy weight in a six-thousand-year-old matrix of interrelated weight standards covering most of the Old World; but it must be admitted that this is of interest chiefly to historians of the subject. The general reader will be forgiven for skipping the rest of this chapter and picking up the thread in Chapter 3.

As we have just seen, the use of troy weight for coinage and medicines extended over the whole of northern Europe, from Britain in the west to Russia in the east, and south down through the old Austro-Hungarian Empire clear to the border with Turkey, the threshold of the middle east. Weight standards based on other ancient originals dominated in France, Spain, and Italy, but on the other side of the Mediterranean, the superior technology of the medieval Arabs, who perfected the art of clearly labeled, accurately made glass weights, brought the troy family of units to a high degree of standardization.[64] The 48-grain unit was the dirhem (half the size of the original drachma from which it got its name); a variety of examples put the medieval dirhem in Damascus at 3.086 to 3.0898 gm (47.62 to 47.68 grains) and the medieval Egyptian dirhem at 3.125 gm (48.225 grains).[65] The French expedition of 1799 found the dirhem weight in use in Cairo at 3.0884 gm (47.66 grains), and the Commission of 1845 at 3.0898 gm (47.68 grains).[66] A unit of about 72 grains, half of the ancient 144-grain qedet, was carried on as the miscal or metical; a variety of medieval examples put this at 4.68 grams (72.2 grains).[67]

Nineteenth- and 20th-century survivals of the Arabic members of the troy family show how assiduously it was maintained over the centuries in some places. Table 2.2, from sources published in the 1880s, tells the story. While the troy connection is most clearly seen in the dirhem and miscal, cognates of the troy ounce, the haverdepois pound and the pound of Cologne are evident as well. Another set of late 19th century survivals (Table 2.3) shows the same units in cities that had adopted the higher standard of Constantinople.[68]

By the mid-20th century, these city standards had all been subsumed by national standards, and in virtually every case had been officially replaced by units of the metric system; but some still stubbornly persisted in popular use, as reported in Table 2.4 by the United Nations.[69] Of the countries listed here, Iran (ancient and medieval Persia) is by far the most important, and the almost exact Persian equivalents to the troy pound and the pound of Cologne are especially significant.

As with the 19th century city standards, there is also a group of late survivals that show the influence of the higher standard of Constantinople (Table 2.5).

The differences between these Arabic cognates prove that they are not just colonial adoptions of the British system but rather independently maintained versions of an ancient common ancestor.

The obvious similarity of the Arabic and English troy weight systems persuaded earlier historians[104] that the English inherited the troy system from the Arabs. Indeed, the example of Charlemagne, who according to tradition actually did inherit the old French weights and measures from Caliph Harun-al-Rashid,[105] makes this not an unreasonable assumption, and it is certainly in keeping with our adoption of Arabic numeration and other large helpings of Arabic science and technology. But the Saxon graves prove that troy weight was in use in England more than a century earlier than the time of Charlemagne, and the Melandra weight (page 26) shows that it was in use during the Roman occupation. The similarity between the

City (1880s)	Unit	Reported weight (gm)	Weight in grains	Troy cognate	Difference
Alexandria[70]	dirhem	3.0884	47.66	48	−0.71%
Algiers[71]	metical	4.66345	71.97	72	−0.04%
	uchia	31.0897	479.79	480	−0.04%
	rotl-feudi	497.435	7676.60	7680	−0.04%
Bet-el-faki[72]	vachia	30.8257	475.71	480	−0.90%
	rottolo	462.385	7135.69	7200	−0.90%
	maund[73]	—	14,273	14,400	−0.90%
Cairo (1914)[74]	dirhem	3.12	48.15	48	+0.31%
	ukia	37.44	577.8	576	+0.31%
Mecca[75]	coffala	3.1030	47.89	48	−0.24%
	miscal	4.6545	71.83	72	−0.24%
	vacheia	31.030	478.87	480	−0.24%
	rattel[76]	498.320	7690.25	7680	+0.13%
Shiraz[77]	miscal	4.6	71	72	−1.4%
Tauris[78]	mascal	4.6488	71.74	72	−0.36%
	dirhem	9.2976	143.48	144	−0.36%
	rottel	464.88	7174.2	7200	−0.36%
Tehran[79]	derhem	3.067	47.33	48	−1.41%
Tripoli[80]	derhem	3.052	47.10	48	−1.91%
	uchia	30.52	471.00	480	−1.91%
Tunis[81]	uchia suki	31.08	479.6	480	−0.08%
Zanzibar[82]	vachiah	28.0668	433.137	432[83]	+0.26%

Table 2.2: 19th century Arabic troy cognates

Chapter 2. The troy pound

City (1880s)	Unit	Reported weight (gm)	Weight in grains	Troy cognate	Difference
Aleppo[84]	dirhem	3.18785	49.20	48	+2.49%
Beirut[85]	dirhem	3.18785	49.20	48	+2.49%
Constantinople[86]	dirhem	3.20259	49.42	48	+2.97%
Damascus[87]	dirhem	3.18785	49.20	48	+2.49%
	metical	4.78178	73.79	72	+2.49%
	vokia	31.8785	491.96	480	+2.49%
Smyrna[88]	dirhem	3.2026	49.42	48	+2.97%

Table 2.3: 19th century Arabic troy cognates (standard of Constantinople)

32

Chapter 2. The troy pound

Country (1960s)	Unit	Reported weight (gm)	Weight in grains	Troy cognate	Difference
Afghanistan[89]	misqal	4.6003	71.0	72	−1.42%
Cyprus[90]	dram	3.175	49.0	48	+2.08%
Egypt[91]	dirhem	3.12	48.15	48	+0.31%
Ethiopia[92]	woket	28	432.1	432	+0.02%
Eritrea[93]	dirhem	(28.1)	433.1	432	+0.25%
Iran[94]	miskal	4.64	71.6	72	−0.56%
	dirhem	9.28	143.2	144	−0.56%
	pinar	92.79	1432	1440	−0.56%
	danar	185.58	2864	2880	−0.56%
	abbassi	371.17	5728	5760	−0.56%
	rottel	463.96	7160	7200	−0.56%
Libya[95]	dirham for silver and silk	3.06748	47.34	48	−1.40%
	ukia for silver and silk	30.6748	473.38	480	−1.40%
Somalia[96]	okia	28.0	432.1	432	+0.02%
Sudan[97]	dirham	3.12	48.15	48	+0.31%
	mithkal[98]	4.68	72.22	72	+0.31%
	okia or wagia	37.44	577.79	576	+0.31%

Table 2.4: 20th century Arabic troy cognates

Chapter 2. The troy pound

Country (1960s)	Unit	Reported weight (gm)	Weight in grains	Troy cognate	Difference
Jordan[99]	dirham	3.2051	49.46	48	+3.05%
Lebanon[100]	dirham	3.205	49.46	48	+3.05%
Libya[101]	ordinary dirham[102]	3.205	49.46	48	+3.05%
Syria[103]	dirham	3.205	49.46	48	+3.05%

Table 2.5: 20th century Arabic troy cognates (standard of Constantinople)

English and Arabic systems, therefore, comes originally not from direct contact but rather common ancestry.

It is very likely, however, that the adoption of Arabic weights by Charlemagne did strongly influence the shape of the English system. Offa, the Anglo-Saxon king of Mercia (A.D. 757–96), visited the court of Charlemagne and then created the first English coinage by so closely copying Arabic originals that his version of the gold dinar actually bore his name and title in Arabic characters.[106]

In the heavier weights (i.e., beyond the units used for gold, silver, and other precious substances), the later Arabic and English systems typically went their separate ways, but the troy cognates sometimes appear again as equivalents of larger decimal multiples of the troy pound (units corresponding to 100 troy pounds or 576,000 grains being the most common) and the larger decimal multiples of the 144-grain qedet, particularly 14,400 and 144,000 grains. The reader will no doubt recognize 576,000 grains as the Aeginetan talent, 14,400 grains as the ancient Egyptian sep, and 144,000 as the next unit up in the typical decimal Egyptian series.

In the system used in Saudi Arabia until 1964, for example, the main commercial unit of weight was the maund of 82.2 pounds avoirdupois (575,400 grains) and the ruba of $\frac{1}{4}$ maund or 20.6 pounds avoirdupois (143,850 grains);[107] these units are (given the stated avoirdupois equivalent) within roundoff error of the corresponding troy units of 576,000 and 144,000 grains, or 100 troy pounds and 25 troy pounds. In the port of Hodeida, in what is now Yemen, the late 19th century weight of the bahar was 374.213936 kg[108] or 577,500.356 troy grains, about 0.3 percent above the troy equivalent. One fortieth of a bahar was a frehsil, at 144,375 grains clearly derived from the Egyptian unit of 10 sep (144,000 grains), and the oka ($\frac{1}{300}$ bahar) was 192,500 grains, which is a thousand ancient Egyptian beqas of 192 grains. The Syrian talent of 432,000 grains can be seen in the Alexandrian kantar, which survived into the mid-20th century in Sudan at a weight of 139.776 kg, which is $5 \times 431,415$ grains, about 0.1 percent below the troy standard.[109] The slight differences between these weight standards and their troy counterparts attest, on the one hand, to the antiquity of the divergence

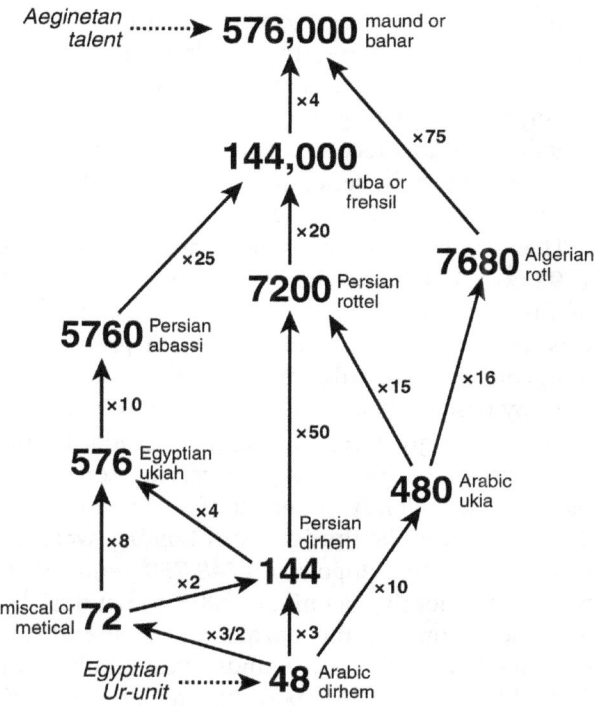

Figure 2.9: 19th and 20th century Arabic troy survivals (compare with Figure 2.8)

between the European and Arabic versions of the standard and the level of care, on the other hand, that went into their independent maintenance over many centuries of separation.

Troy weight in India

The best example of the troy standard in the east is the premetric weight system of India and Pakistan. Here the exact official alignment with English troy weight makes for a nice demonstration of this variant tradition:

1 maund = 576,000 grains (100 troy pounds)

1 seer = $\frac{1}{40}$ maund = 14,400 grains (the ancient Egyptian sep)

1 pao = $\frac{1}{4}$ seer = 3600 grains (the mark of Cologne)

The smaller units were (and are) peculiar to India; the most important was the tola of $\frac{1}{80}$ seer or

Chapter 2. The troy pound

180 grains. The exact correspondence with troy weight was, of course, the result of a conscious shift to British Imperial standards that occurred relatively late, beginning around 1833, but the adjustment needed to effect this change in basis was quite small; compare the series above with the old bazaar weights of Calcutta in Table 2.6.[110]

The pre-1833 sicca weight at Calcutta was 179.666 troy grains, whereas the pre-1833 weight of the rupee coined by the mint at Ferruckabad was 180.234 grains;[111] both of these primary Indian coinage standards were within 0.2 percent of the troy-based tola later defined by the British.[112]

Two other great regional Indian standards survived into the 19th century, preserving some very old units just slightly under the troy standard. The Surat khandi of the Presidency of Bombay weighed $821\frac{1}{4}$ pounds avoirdupois or 5,748,750 grains (0.20 percent smaller than a unit of 1000 troy pounds),[113] and the maund of the bazaar of the Presidency of Bengal was 82.1392 pounds avoirdupois[114] or 574,974.4 grains (0.18 percent smaller than 100 troy pounds). Like the official Imperial maund and the pre-metric maund of Calcutta, all of these heavier weights are survivals of the Aeginetan talent (or a unit of 10 talents in the case of the Surat khandi of Bombay).

The large old troy-cognate units in Saudi Arabia, India, and Pakistan stand in contrast to the much smaller coinage weights typical of troy measure in most of the rest of the world. Their size suggests that the ancient ancestor of these standards was originally used for all purposes of trade—in other words, that its restricted application to precious metals, perfumes, medicines, spices, and other fine wares in England and America was a later development.

Troy weight in the Far East

Continuing on to the Far East, we find a decimally divided troy pound, a little above the western standards but now back in character as a coinage weight, particularly for silver.

In most of eastern Asia, the standard for precision weighing was a unit called the tael or liang that corresponded, with slight variations, to the Anglo-

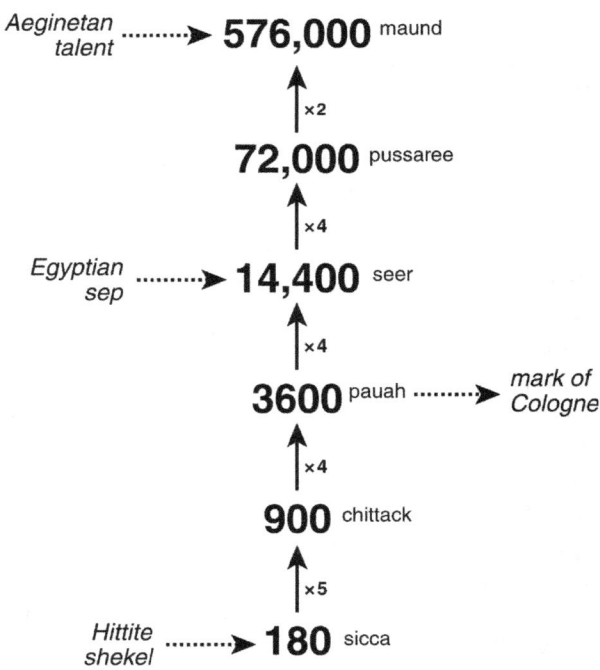

Figure 2.10: Troy weight in India

Saxon Grove Ferry weight of 576 grains (i.e., $\frac{1}{10}$ troy pound). The oriental systems of weight were almost completely decimal; had they been constructed with the uniformity of the metric system, the next unit up from the tael would have corresponded directly to the troy pound of 5760 grains. For practical purposes, however, all the major far eastern systems inserted a strong binary element at this point, multiplying the tael by 16 rather than 10 to form the next larger unit, known in Chinese as the kin or more generically as the catty. All the larger weights were decimal multiples of the catty (corresponding to a unit of 9216 grains in the troy system), and all the smaller ones were decimal submultiples of the tael. The first subunit of the tael was the ch'ien or mace, $\frac{1}{10}$ of a tael; this corresponds to the same troy unit of 57.6 grains found in the Saxon graves and in the pharmaceutical systems of Germany and Russia.

Judging from 19th- and 20th-century survivals, there appear to have been two distinct versions of the tael, a true troy standard weighing between 575 and 576 troy grains and a slightly heavier version weighing around 583 troy grains.[115]

Unit	Reported wt. (gm)	Weight in grains	Troy cognate	Difference
maund	37,255.075	574,933.66	576,000	−0.19%
pussaree	4656.884	71,866.71	72,000	−0.19%
seer	931.377	14,373.34	14,400	−0.19%
pauah	232.844	3593.34	3600	−0.19%
chittack	58.211	898.33	900	−0.19%
sicca	11.642	179.67	180	−0.19%

Table 2.6: Pre-Imperial bazaar weights of Calcutta

The lower version, corresponding most directly to troy weight, is exemplified by the Kuping (Treasury) tael of 575.94 grains, used until the beginning of the 20th century for the payment of most government debts other than customs duties;[116] this is almost exactly the tenth of a troy pound. As late as the 1920s the Chinese government was attempting to achieve national standardization on a tael of 37.301 grams or 575.64 grains, barely 0.06 percent off the troy standard.[117]

In medieval Sumatra and Java we find older forms of the Asian system showing this lower version and its relationship with troy very clearly. Srivijayan coins (south Sumatra, 7th to 14th century) point to a chang or kati of 1127 grams, which is 3 troy pounds of 5797.4 troy grains, about 0.6 percent above our version. And the weight of the cati or kin of Java in 1416 has been estimated at 1120 grams, which is 3 pounds of 5761.4 troy grains—almost identical to the troy pound.[118]

For ordinary commerce, however, the heavier tael of about 583 grains predominated, and this was the version generally used in the surrounding countries that used the Asian system. For the convenience of occidentals used to dealing in avoirdupois pounds, the commercial or Haikwan tael was defined by treaty as $\frac{4}{3}$ of a modern avoirdupois ounce or $583\frac{1}{3}$ troy grains, making the catty of 16 taels exactly equal to $\frac{4}{3}$ of an avoirdupois pound; at this value, the tael continued to be used well into the 20th century in China, Hong Kong, Macau, Malaya, Singapore, Borneo, and Sarawak.

While the value for the tael of exactly $\frac{4}{3}$ of an avoirdupois ounce was obviously arrived at to suit users of the British Imperial system, equivalents of the same units defined by treaty with countries using the metric system show that the original unit must have been fairly close to this value; by treaty with France and Italy, the catty was 604.53 grams,[119] putting the tael at 583.08 troy grains.

The premetric systems of several other far eastern countries show a similar standard: in Siam (now Thailand) as late as 1891 the pecul (100 catties) was 133.35 pounds,[120] corresponding to a tael of 583.4 troy grains; in North Vietnam the can (catty) of 604.50 grams, corresponding to a luang (tael) of 583.054 troy grains, continued to be used into the 1950s;[121] and as far east as the Sulu Archipelago in the Philippine Islands, the tael was used into the 1890s at 37.7849 grams or 583.110 troy grains.[122] In countries far from the centers of trade the value of the tael drifted even higher, to 37.8529 grams or 584.160 troy grains in Batavia (now Jakarta, capital of Indonesia)[123] and 590.75 troy grains in Cochin China (now part of South Vietnam);[124] but the bulk of the evidence shows these higher values to be aberrations.

Between the two fairly well-defined groups of tael standards at 574–76 troy grains and 583–84 troy grains there was a third variety that may have been formed by a later attempt at a compromise between these two values. In Canton and Shanghai (China) the tael for transactions in foreign bar silver was 579.85 troy grains,[125] and something like this unit served as the basis for the traditional weight standard of Japan and Korea; the official value of the premetric Japanese kwan was $\frac{3125}{378}$ avoirdupois pounds or 57,870.4 troy grains,[126] putting the yang (liang or tael) of $\frac{1}{100}$ kwan at 578.704 troy grains. There was also a Japanese unit of 10 yang called the hiyaka-me that corresponded equally closely to the troy pound itself.[127]

This compromise tael is almost exactly 37.5 grams, putting the catty at a very handy 600 grams; and in the 20th century these became the official metricized values for the tael and catty equivalents in Laos, Cambodia, Thailand, and Vietnam as well as Japan and Korea. In Laos, the pong for opium remains, at a metricized value of 375 grams (5787.13 troy grains), the last oriental vestige of the troy pound itself.[128] Its specialized use is one that the troy standard always served in our own tradition: the sale of drugs.

Notes

[1] *Report from the Select Committee* (1821), 5.

[2] Ferguson, "The Carlisle Bushel," 306, citing *Journal of the British Archaeological Association* 8:370.

[3] This apparently became a common pattern for troy weights. An 1821 law of the State of Maine requires the Treasurer of the state to "cause to be had and preserved as public standards ... a nest of Troy weights from one hundred and twenty eight pounds, down to the least denomination" (*Laws of the State of Maine* (1821), 2:575–76). The edition of 1847 shows that *pounds* is a misprint for *ounces* (*Revised Statutes of the State of Maine* (1847), 2:320.).

[4] [Chisolm], *Seventh Annual Report*, 21. The engraving is from Chisolm, *On the Science of Weighing and Measuring*, 61.

[5] That is, any system in which all the standard weights are put on one side of the balance against the thing being weighed. The systems you get when you allow some of the standards to be put in the other pan along with the thing being weighed, effectively performing a subtractive function, are extremely interesting and will be the subject of a future study that explains, among other things, why many unmarked ancient weights should be interpreted as multiples of 3, 9, and 27.

[6] Chisolm, *On the Science of Weighing and Measuring*, 69.

[7] "It was impossible to avoid remarking, that the weight hitherto adopted for the standard weight of this empire, viz., the troy pound, is comparatively useless even in the few trades or professions in which troy weight is commonly used, and that to the great mass of the British population it is wholly unknown." *Report of the Commissioners*, 1841, 4. Note that the reference is to the troy pound, not the troy ounce, which remained in common use for bullion. According to Chisolm (writing in *Nature* 8:308), the Imperial pound avoirdupois became the legal standard of weight in Britain in 1855.

[8] The story of the creation of the troy pound of the United States and its subsequent history is from Judson, *Weights and Measures Standards of the United States*, 5 and 27. The photo of the troy pound is from page 4 of this publication.

[9] According to Eleanor McKelvey, Curator of Exhibits at the U.S. Mint in Philadelphia (Karma A. Beal, Historical Information Specialist, Information Resources and Services Division, National Institute of Standards and Technology; personal communication, 4 January 1991).

[10] The typical 325 mg dosage for over-the-counter pain medications is the metric equivalent of 5 grains (323.99 mg) rounded up to the nearest 5 milligrams. The 81 mg daily adult dose of aspirin often prescribed now for heart health is simply the old children's dose, $1\frac{1}{4}$ grains, expressed as an ungainly metric equivalent.

[11] CONNOR, 2.

[12] CONNOR, 4.

[13] I found the pennyweight still in use among the wholesale jewelers of Manhattan's 47th street as late as the 1990s.

[14] These facts are too well established to need attribution. The reader wishing to explore the subject further would be well served by Skinner's *Weights and Measures*, which remains the best general introduction to the subject.

[15] Latin *uncia* is also where we get the word *inch*. Readers wanting to know more about the etymology of Customary unit names will find the Glossary beginning on page 139 of interest.

[16] Landels, *Engineering in the Ancient World*, 171.

[17] Thureau-Dangin, "Sketch of a History of the Sexagesimal System," 110.

[18] Petrie, *Prehistoric Egypt*, 28–29; *Ancient Weights and Measures*, 14.

[19] The identification of this unit is generally credited to Flinders Petrie, who based the finding on thousands of weights excavated and painstakingly reconstructed in his researches of the late 19th and early 20th centuries. The value of 144 grains for the commercial shekel has been generally recognized for over a century, though it is now always given as 9.33 grams. SKINNER notes the connection between this unit and the troy system in several places.

[20] SKINNER, 3. There was a U.S. coin corresponding to this weight, though you have to be getting along in years to remember it; until it stopped being made of silver, the U.S. half dollar coin weighed 192 grains, exactly.

[21] Petrie, *Ancient Weights and Measures*, 18, noting "the frequency of 40 qedet weights (= 30 beqa) and $\frac{1}{3}$ qedet (= $\frac{1}{4}$ beqa) in later times," i.e., later than the Old Kingdom.

[22] Bass, "Cape Gelidonya," 135–42. The categorization shown here is my own, but it is substantially in line with Bass's.

[23] Pulak, "The balance weights from the Late Bronze Age shipwreck at Uluburun," 261. These are Pulak's findings, and it is worth noting that the categorization of the weights representing the qedet standard was done by computer analysis rather than through human interpretation.

[24] Bible readers will recognize Carchemish as the site of a famous battle between the Egyptians and the Babylonians that took place somewhat later, around 605 B.C. (Jeremiah 46:2 and 2 Chronicles 35:20). Carchemish is also mentioned in Isaiah 10:9.

[25] 1 Samuel 13:19–22.

[26] Michailidou, *Weight and Value in Pre-coinage Societies*, 20; Parise, "Mina di Ugarit, mina di Karkemish, mina di Khatti," 156, 158; Parise, "Unita ponderali e rapporti di cambio nella Siria del Nord," 129.

[27] *Encyclopædia Britannica,* 11th ed., s.v. "Pheidon."

[28] Hitzl, *Gewichte griechischer Zeit aus Olympia,* 111. I am grateful to Professor Ronald S. Stroud of U.C. Berkeley for calling this study to my attention.

[29] This "stater" should not be confused with the much smaller Greek coin unit of that name (equivalent to a didrachm). It appears to have been common in many places to use the name of a larger unit for coins that are some well-known but unstated subdivision of that unit, and indeed, Pheidon himself coined a silver stater (coin) of $\frac{1}{60}$ stater (weight), i.e., 192 grains (SKINNER, 58).

[30] SKINNER, 59, 86.

[31] Hussey, *Essay on the Ancient Weights and Money,* 60. Hitzl (*Gewichte griechischer Zeit aus Olympia,* 53) calculates the weight of the Aeginetan drachma as 6.237 grams (96.25 grains).

[32] While commonly cited, the supposed derivation from a medieval name for London seems particularly unlikely given the use of the word *troy* to refer to their native weights by both the Scotch and the Dutch.

[33] Petrie, *Ancient Weights and Measures,* 43 and pl. 49. The mixture of different standards is typical of all randomly assembled collections of ancient weights from a given location.

[34] Britain is believed to be the "Tin Islands" described by Herodotus (5th century B.C.). The tin trade in Britain, which apparently began with the Phoenicians and Carthaginians, perhaps as far back as 1500 B.C., was described by the the Greek historian Polybius (2d century B.C.) and the Sicilian historian Diodorus Siculus (1st century B.C.). See Evans, *Popular History of the Ancient Britons,* 2–4, and Lubbock, *Pre-historic Times,* 57–71.

[35] Petrie, *Ancient Weights and Measures,* 28.

[36] MARTINI, 144 (pfund = 357.78 grams; $\frac{1}{12}$ pfund = unze, $\frac{1}{8}$ unze = drachme). CLARKE (35) gives the medicinal pfund as 0.788974 lb av, which would make the drachme 57.5294 grains.

[37] MARTINI, 524–25 (medicinal funt = 358.322615 gm, unzia = 29.860218 gm, drachma = 3.732527 gm).

[38] See SKINNER 84–87 for a discussion of late classical and early medieval weight standards and their antecedants.

[39] This statute was not the first to refer to troy weight by name; that distinction belongs to 2 Henry 5 stat. 2 c. 4 (1414) regulating the gilding of silver ware and later 2 Henry 6 c. 16 (1423) regulating the price of a pound of silver.

[40] [Chisolm], *Seventh Annual Report,* 15. American readers need to remember that in British references earlier than the 20th century, "corn" just means "grain," usually wheat.

[41] Unless otherwise noted, all the Saxon and Irish/Viking weights cited here are from Smith, "Early Anglo-Saxon Weights," 122–129).

[42] The figure given here excludes the fractional weights as more subject to error and includes 10 weights that could alternatively be ascribed to a unit of 50 grains. If these slightly heavier weights are excluded, the remaining 20 average 47.74 troy grains. Weights of several other standards are also found in the graves, which is consistent with the trading environment and typical of most collections of ancient weights; some of them belong to the tower standard discussed in Chapter 4, and most of the remainder belong to Norse and Roman coin standards.

[43] Brent, "Anglo-Saxon Cemetery at Sarr," 157–63.

[44] *Proceedings of the Society of Antiquaries of London,* 2nd ser., 16 (1895–97): 174–75.

[45] Smith, "Early Anglo-Saxon Weights," 122.

[46] Smith, "Early Anglo-Saxon Weights," 127, citing W. Airy, *Minutes of Proc. Inst. of Civil Engineers,* 191 Part 1 (1912–13).

[47] Also discussed in SKINNER, 90.

[48] Hall and Nicholas, *Tracts and Table Books,* 12–13.

[49] Though the "h" continued to be pronounced into the 18th century; the OED cites a quotation from 1701: "*h* may be sounded in halleluiah, habiliment, hauer-du-pois, etc."

[50] *Report from the Select Committee of the House of Lords* (1823), 12.

[51] MARTINI, 169.

[52] MARTINI, 184. CLARKE cites a Norwegian unit called the skålpund weighing 0.9378 lb av or 6565 grains with an untz of 410.3 grains; this might be a Roman survival.

[53] MARTINI, 99. JOHNSON'S, 486, and CLARKE, 20, both give this as 1.09906 lb av (7693.42 grains).

[54] Stratmann, *Middle-English Dictionary,* 567.

[55] Harris, *Coventry Leet Book,* part 2, 396.

[56] [Chisolm], *Seventh Annual Report,* 29.

[57] Wright, *English Dialect Dictionary,* 5:661.

[58] The OED provides an example from 1809: "He goes out almost every week to eat speck with the country folks; thereby showing that a democratic governor is not to be choaked with fat pork."

[59] *Report from the Committee* (1758), 67; see also 69.

[60] In Germany and Scandinavia, this binary division continued downward; pound $\times \frac{1}{2}$ = mark or marc (for coinage) $\times \frac{1}{2}$ = unze or once (same as the troy ounce) $\times \frac{1}{8}$ = quentchen (Danish quintin, corresponding to our old apothecaries' dram of 60 troy grains) $\times \frac{1}{4}$ = ort.

[61] Kisch, *Scales and Weights,* 9. The weights are 467.548 gm, 467.737 gm, and 468.125 gm.

[62] CLARKE, 62 (pfund = 1.03118 lb av).

[63] Miller, "On the Construction of the New Imperial Standard Pound," 755.

[64] Glass is a wonderful material for making weights because it wears well and doesn't corrode. English tradesmen were still using glass weights to test gold coins as late as the 1880s (Lucas, "On 'The British Standards'," 36).

[65] Hinz, *Islamische Masse und Gewichte,* 3.

[66] Zambaur, "Dirham," 979.

[67] Hinz, *Islamische Masse und Gewichte,* 3.

[68] This higher subset may possibly be related to the cluster at about 50 grains noted above for the weights found in Saxon and Viking graves.

[69] *World Weights and Measures: Handbook for Statisticians* (UN, 1966). An earlier edition of this work appeared in 1955; all citations of "UN" below are to the 1966 edition unless otherwise noted. Values for all of the weights in the UN handbooks are reported in both metric and British units; the metric units are reasonably uniform, but the British equivalents are given in a mixture of grains, troy ounces, avoirdupois pounds, avoirdupois ounces, and avoirdupois drams that contains so many roundoff errors that consistent comparisons are almost impossible. Unless otherwise noted, therefore, all the troy equivalents shown here are calculated from the stated metric values.

[70] MARTINI, 20.

[71] MARTINI, 20. CLARKE (8) gives the "rotl feuddi, for gold and silver" as 1.096714 lb av (7677 grains) and the metical as 72.06 grains.

NOTES

[72] MARTINI, 85 (except for the maund; see next). CLARKE (10) reports the rotl at 1.019531 lb av (7136.717 grains).

[73] CLARKE, 10 (maund is 2.039 lb av).

[74] Set by law in 1914 (TATE'S, 182–84). CLARKE (29) gives its value in 1877 as 47.73 grains. An 1886 committee agreed on a value of 3.0898 gm (47.683 grains); 3.12 gm (48.15 grains) was recommended in 1891, and this was the value eventually adopted (Curry, *The Control of Weights and Measures in Egypt*, 1–2).

[75] MARTINI, 341. The same standard was used in Mocha (369).

[76] The commercial weight, which was divided into 15 vacheias of 33.221 gm. The three weights for gold and silver listed above are based on a different vacheia.

[77] MARTINI, 735.

[78] JOHNSON'S, 487; the rottel is given as 1.027 lb av, which would be 465.84 grams or 7189 grains.

[79] MARTINI, 770.

[80] "Tripoli di Barberia (Reggenza di Tripoli)." MARTINI, 797.

[81] MARTINI, 802. The uchia suki was used for meat, oil, soap, butter, etc.

[82] MARTINI, 841.

[83] That is, 3×144.

[84] MARTINI, 19.

[85] MARTINI, 65.

[86] MARTINI, 179.

[87] MARTINI, 190.

[88] MARTINI, 742.

[89] UN, 31.

[90] UN, 47.

[91] In legal force until 1956. UN (1955), 58.

[92] UN, 50.

[93] UN (1955), 60. The value of the Eritrean okia of 16 dirhems is reported as 449 gm, and the value of the dirhem is calculated from that.

[94] UN, (1966) 60; UN (1955), 75. Roundoff errors in the figures provided make it impossible to establish the base value used by the UN statisticians with certainty, so here I've taken the 1955 metric value of the rottel (463.96 gm; given as 463.9 gm in 1966) as definitive and calculated both metric and troy values from that base. The almost exact equivalence of 463.96 gm and 7160 troy grains makes it likely that the latter value was the base upon which the rest of the UN figures for Iran were calculated.

[95] UN (1955), 86. The 1966 edition (67) gives the 1955 edition as the source but reports the ukia for silver and silk as 30.672 gm.

[96] UN, 88.

[97] UN, 91.

[98] UN (1955 edition), 28.

[99] UN, 63.

[100] UN, 66.

[101] UN, 67.

[102] Compare this with the value of the Libyan weights for silver and silk in Table 2.4.

[103] UN, 92.

[104] For example, [Chisolm], *Seventh Annual Report*, 31.

[105] SKINNER, 88–89.

[106] SKINNER, 89.

[107] UN, 85. The metric value for the maund is given as 37.285 kg, which would make the maund 575,395 grains.

[108] MARTINI, 253.

[109] UN (1955), 28.

[110] MARTINI, 130 (maund = 37.255075 kg).

[111] Martin, *History of the Colonies of the British Empire*, 342; Prinsep, "On the Standard Weights of England and India," 445.

[112] The pre-1833 weight of the Calcutta sicca rupee was 191.916 grains—the ancient Egyptian beqa again (Martin, *History of the Colonies of the British Empire*, 343; Prinsep, "On the Standard Weights of England and India," 445).

[113] Murray, *Handbook of the Bombay Presidency*, 40.

[114] JOHNSON'S, 486.

[115] There were also a number of taels corresponding to an unrelated unit of 540 to 560 grains. A traditional account of the relationship between Chinese units of weight, length, capacity, and musical pitch (!) based on the size of millet seeds can be found in an unattributed article titled "The Connection Between Chinese Music, Weights, and Measures" (*Nature* 30 (Oct. 9, 1884): 565–66) and in a follow-up letter by J. P. O'Reilly under the heading "On Two Jade-handled Brushes" (*Nature* 35 (Feb. 3, 1887): 318–19).

[116] Hoang, *Notions techniques*, 55, citing the value "employé pour le trésor public"; but "quand on envoie l'argent au trésorier métropolitain, 1 liang, égale 37 gram. 417," i.e., 577.43 grains. TATE'S, 237, puts the value of the Kuping tael at 575.8 grains.

[117] TATE'S, 240–41. The UN *Glossary of Commodity Terms* (p. 97) reports this value still in use as late as 1954.

[118] Gear and Gear, *An Ancient Bird-Shaped Weight System from Lan Na and Burma*, 10–11.

[119] MARTINI, 515; Hoang, *Notions Techniques*, 56. TATE'S, 241, gives the catty by treaty with France at 604.5 grams.

[120] CLARKE, 69.

[121] UN (1955), 143. This weight, also listed in the 1954 UN *Glossary of Commodity Terms* (106), had disappeared by the 1966 edition of the Handbook.

[122] CLARKE, 115.

[123] CLARKE, 114.

[124] CLARKE, 24.

[125] TATE'S, 232–33.

[126] Imperial Japanese Ordinance No. 169 (1 pound = $\frac{378}{3125}$ kwan). *Report by the Board of Trade* (1910), 52.

[127] *Encyclopædia Britannica*, 11th Edition, s.v. "Weights and Measures."

[128] UN, 66.

Chapter 3
The wine gallon

The Customary System of liquid measures is based on a standard historically known as the wine gallon. The first English settlers in the U.S. brought over this standard and its associated system, and it has remained the U.S. standard ever since. The gallon system is quite elegant, and the wine gallon standard itself is very ancient, having been maintained within one percent of its original value for over two thousand years.

Notice the distinction I'm making here between gallon *system* and gallon *standard*. *System* means the way in which the units relate to each other—for example, the fact that there are 8 pints in a gallon. *Standard* refers to the actual size of the base unit, which in the case of the U.S. gallon is 231 cubic inches (hereafter generally abbreviated in^3). In medieval England, there was just one gallon *system*, but there were at least four gallon *standards*—the Guildhall gallon, the wine gallon, the ale gallon, and the Winchester gallon.

In the great reorganization of the British system after the destruction of the Parliamentary standards described in Chapter 2, all these gallon standards were discarded in favor of a brand new, synthetic standard called the Imperial gallon. The Imperial gallon was designed to provide the same kind of relation between capacities and weights offered by the recently invented metric system—in other words, to show up the French—and it therefore defined a gallon as the measure holding 10 avoirdupois pounds of water, as the French system defined a liter as the measure holding a kilogram of water.[1] Just as we continued basing our weights on our own copy of the troy pound, however, here in the U.S. we kept the best established and longest used of all the old gallon standards, and the first to be defined by English statute in terms of cubic inches—the wine gallon. But let's briefly review the

Figure 3.1: Exchequer standard wine gallon of Queen Anne

system before we start talking about the standard.

The customary gallon system

The characteristic feature of our traditional gallon system is its thoroughgoing dependence on the number 2 (or $\frac{1}{2}$). Most of us are familiar with the binary subdivisions of the gallon, even if we no longer remember all of their names; a complete list is given in Table 3.1.

The teaspoon was originally identical to the fluid dram ($\frac{1}{4}$ tablespoon); see Figure 3.2. Over the course of time, however, the teaspoons actually used in American table settings grew somewhat larger than this, and in the early part of the 20th century, the U.S. Department of Commerce finally redefined the teaspoon as $\frac{1}{3}$ tablespoon.[2] The minim ($\frac{1}{480}$ fluid ounce) is roughly the volume of a

Chapter 3. The wine gallon

Figure 3.2: Early 20th century American glass medicine cups showing former definition of the teaspoon as a quarter of a tablespoon, equivalent to the fluid dram ($\frac{1}{8}$ fluid ounce). The teaspoon is now defined in the U.S. as a third of a tablespoon ($\frac{1}{6}$ fluid ounce)

	Unit	Gallons	Fluid Ounces
	gallon	1	128
$\times \frac{1}{2} =$	pottle	$\frac{1}{2}$	64
$\times \frac{1}{2} =$	quart	$\frac{1}{4}$	32
$\times \frac{1}{2} =$	pint	$\frac{1}{8}$	16
$\times \frac{1}{2} =$	half pint (cup)	$\frac{1}{16}$	8
$\times \frac{1}{2} =$	gill	$\frac{1}{32}$	4
$\times \frac{1}{2} =$	wine glass	$\frac{1}{64}$	2
$\times \frac{1}{2} =$	fluid ounce	$\frac{1}{128}$	1
$\times \frac{1}{2} =$	tablespoon	$\frac{1}{256}$	$\frac{1}{2}$
$\times \frac{1}{4} =$	fluid dram	$\frac{1}{1024}$	$\frac{1}{8}$
$\times \frac{1}{60} =$	minim	$\frac{1}{61440}$	$\frac{1}{480}$

Table 3.1: The Customary gallon (wine gallon) system

drop of water; it still appears occasionally in niche applications such as specifying the volume of capsules sold in health food stores.

Not all of these unit names were ever in use at any one instant in history. The cup, for example, as the name for a specific subdivision of the gallon, is of quite recent origin; historically and legally, that unit is called a half pint. But most of these names were in use at every period of the Anglo-American tradition in capacity measures.

Our collective amnesia about the names of some of these units has done little to diminish our actual use of them. For example, hardly anyone these days knows that a half-gallon used to be called a pottle, though the name appears in state laws as late as 1803,[3] but the measure itself continues to thrive on supermarket shelves and has recently been resurrected for beer and renamed the *growler*. Likewise the gill (pronounced "jill," 4 fluid ounces) and the

Chapter 3. The wine gallon

 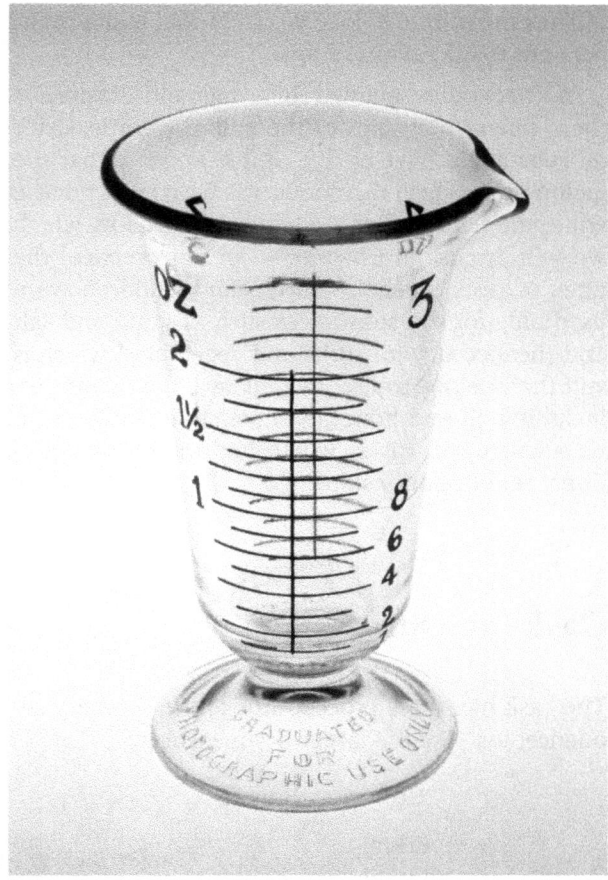

Figure 3.3: *Left:* Late 19th or early 20th century medicine cup showing the wine glass as a unit of measure equal to 2 fluid ounces; *Right:* Late 19th or early 20th century photographic graduate marked in fluid ounces and fluid drams

wine glass (2 fluid ounces) remain common units for the sale of toiletries and cosmetics.

In many parts of England, the gill was equal to a half pint (what we now call a cup in the U.S.) and the measure of a quarter pint was called a jack.[4] "Jack and gill (Jill)" appear in at least one old drinking song and are obviously connected somehow with the names in the nursery rhyme. According to some authors, the line "Jack fell down and broke his crown" refers to the political defeat of an English king who unsuccessfully tried to reduce the size of these tavern measures in order to increase liquor taxes, which were levied by the glass.

A bushel and a peck

As you just saw, our traditional subdivisions of the gallon are almost entirely binary. Another series of binary relationships extends the gallon system upwards:

	Unit	Gallons
	gallon	1
×2 =	peck	2
×2 =	demibushel	4
×2 =	bushel	8
×8 =	quarter	64
×4 =	chaldron	256

Chapter 3. The wine gallon

Thus the refrain "I love you, a bushel and a peck" nets out to 10 gallons of love.

In theory—as implied by some old statutes—these binary multiples of the gallon were used for all substances, wet or dry. But it appears that the gallon upon which this series was based was not the wine gallon but rather the Winchester gallon, which we will discuss in Chapter 4, and in practice, the units of peck, bushel, quarter, and chaldron were used only for dry substances such as grain and salt and (heaped up) for fruits and vegetables, which is still the practice today. Liquids in large quantities, including oil and honey, and meats, including fish, were put up in casks, which had their own traditional set of relationships.

Cask measures

The cask measures proceed upward in a binary sequence, just like the larger dry measures.

	Unit	Barrels
	firkin	$\frac{1}{4}$
×2 =	kilderkin	$\frac{1}{2}$
×2 =	barrel	1
×2 =	hogshead	2
×2 =	pipe or butt	4
×2 =	tun	8

As you may have guessed, "tun" is just a variant spelling of "ton," and in fact a tun of water or wine weighs about a ton.

Notice that I've defined all of these relationships in terms of the barrel rather than the gallon. Unfortunately for the integrity of the old system, the barrel itself was defined as containing different numbers of gallons depending on what it held—30 or $31\frac{1}{2}$ or 32 or 36 gallons to the barrel, depending on whether you were packing beer or salt pork or herrings or butter or whatnot. In the U.S., however, there's only ever been one predominant system of cask measures for liquids, the one traditionally used in England for wine and embedded in U.S. law by the old English statute 2 Henry 6 c. 14 (1423):

	Unit	Barrels	Wine gallons
	barrel	1	$31\frac{1}{2}$
×2 =	hogshead	2	63
×2 =	pipe or butt	4	126
×2 =	tun	8	252

There are also two important auxiliary cask units based on thirds: one called the tertian or (sometimes) puncheon, which is one third of a tun (84 gallons), and a unit half this size, the tierce, which is one third of a pipe (42 gallons). These definitions are all still in force in the U.S. where not specifically superseded by state laws.

The oldest reference to the distinctive $31\frac{1}{2}$-gallon barrel of the wine cask system in U.S. statutes appears in a New York law of 1675 containing the sentence

> The Oyle Cask or Barrells are to contain thirty-one Gallons and a halfe.[5]

A 1685 West Jersey law required

> that all barrels that shall be made after the publication hereof, shall contain one and thirty gallons, and an half at the least, with the cooper's mark thereupon...[6]

The oldest U.S. statute mention of the entire series appears in a Massachusetts Bay Province law of 1692–93 requiring that

> all sorts and kinds of tight cask used for any liquor, fish, beef, pork or any other commodities within this, their majesties' province, shall be of London assize; that is to say, butts to contain one hundred and twenty-six gallons; puncheons, eighty-four gallons; hogsheads, sixty-three gallons; tearses, forty-two gallons; barrels, thirty-one gallons and a half; and made of sound, well-seasoned timber, and free of sap.[7]

This is the classic assize of London. Not all colonial laws followed this series consistently; a 1693 Pennsylvania law required

> that all Tight Cask for Beer, Ales, Cider, pork, beef & oil, and all such Comodities shall be made of good sound, well seasoned white oak timber, and to Contain as followeth—viz: The

Puncheon, Eighty four gallons. Hogshead sixty-five gallons. Tierce fourty two gallons. Barrels, thirty one and a half gallons. Half barrels, sixteen gallons wine measure, according to the practice of our neighboring colonies.[8]

Several of the colonies did indeed round off the half barrel of $15\frac{3}{4}$ gallons to 16, but the reference to a 65-gallon hogshead appears to be unique (and probably just an error).

An act passed by the New York General Assembly in 1703 required

All Tite Barrels to contain Thirty one Gallons and an half of Wine-Measure each, and not to exceed or be half a Gallon over or under the same...[9]

The same wording appears in a Royal Colony of New Jersey Act of 1725.[10] A 1715 law of the Province of North Carolina said that

no Cooper or other person whatsoever making Casks shall expose to sale any Barrels or Halfe Barrels for the holding of Beefe, Pork, Pitch, Tar or Train Oyle but what shall contain & hold Thirty One Gallons & a halfe each Barrel, & Fifteen & three quarters Gallons Each Halfe Barrel...[11]

A 1727 Pennsylvania law put the barrel for beef and pork at "thirty-one gallons and a half, wine measure."[12]

The $31\frac{1}{2}$-gallon barrel of the wine cask system remains in use in many states as the common beer barrel, though about a third of the states[13] have rounded the capacity of the barrel for beer, ale, and other "malt beverages" or "fermented liquors" from $31\frac{1}{2}$ gallons down to 31.[14] The tierce of 42 U.S. gallons ($\frac{1}{3}$ of a pipe or $\frac{1}{6}$ of a tun) is even more widely used; because the early U.S. petroleum industry employed old wine barrels of this size to transport crude, the tierce lives on internationally as "the barrel of oil" whose price we hear about daily on the news.

Oldest statute definition of the wine gallon (1496)

The oldest statute definition of our Customary unit of capacity for liquids is found in 12 Henry 7 c. 5 (1496), mentioned in Chapter 2 as the definitive old English statute relating to troy weight. The passage quoted there is actually the closing part of this larger set of definitions:

the Kinge our Sovereign Lord by the assent of his Lordis spirituall and temporall and Comens in this present parliament assembled and by auctorite of the same, ordeynneth establisseth and enacteth, that the mesure of the busshell conteyn viij galons of whete, and that every galon conteyn viij li. [8 pounds] of whete of troi weight, and every li. conteyne xij unces of troy weight, and every unce conteyn xx sterlinges [pennyweights], and every sterling be of the weight of xxxij cornes of whete that grewe in the myddes of the Eare of the whete according to the old Lawes of this Land. And that it pleaseth the Kingis Highnes to make a standard of a busshell and a galon aftir the seid assise to remayne in his seid Tresory for ever...

Referenced explicitly in a Virginia statute of 1661,[15] this foundational definition was law in all of the American colonies and is still law in the U.S. wherever it has not been superseded by state law (see Appendix B).

The most noticable aspect of the earliest statute definition of our gallon, and perhaps the most surprising to the modern reader, is its definition of capacity in terms of weight. Nowadays we associate this kind of definition with the metric system and its equation of the liter with a kilogram of water,[16] but in the old days, the interesting correspondences were with the weight of the substance that was actually being measured by the gallon, and water was not a significant trade good. People loading a ship with wheat or barley didn't weigh it; up till the 19th century they used capacity measures such as the bushel to fling measured amounts of grain into the hold as quickly as possible. But the master of the ship needed to know the weights being added, because that's what determines the ship's displacement and handling (and this is why a relationship

Chapter 3. The wine gallon

between capacity and weight of water often did enter the picture even though water itself was not being traded). All major ancient systems of measures featured a direct connection between capacities and weights as an aid to calculation, and the old English system was no exception.

The reference to grains of wheat is ritual; in fact, there is no evidence that actual grains of wheat were ever used to weigh anything in any era after the bronze age. Real grains of wheat vary too much in size to function as a natural standard, and while the weight of a grain of wheat implied by the statute ($\frac{3}{4}$ of a troy grain) is not an unreasonable one for real wheat, exactly the same formula was used in connection with what is clearly a different standard of weight, as we'll see in Chapter 4. On the other hand, the *weight per gallon* of good quality wheat is quite stable when averaged over many weighings and seasons,[17] and this average would be the relationship of practical interest to the merchant and to the captain of a ship or barge.

The weight and capacity system described in the statute of 1496 can be reduced to a single elegant relationship:

1 pint of wheat weighs 1 troy pound

This definition is not only wonderfully simple, it is also just about perfect for real wheat. Wheat that weighed one troy pound per U.S. pint would weigh 61.28 avoirdupois pounds per U.S. bushel (Chapter 4); actual U.S. No. 1 wheat weighs about 61.0 pounds per bushel.[18]

Queen Anne defines the wine gallon (1706)

Notwithstanding the simplicity of the original wine gallon system and its ancient history (more on this later), the standard finally fell into dispute in 1688 when His Majesty's Commissioners of the Excise were informed "that the true standard gallon contains only 224 cubic inches, and not 231 cubic inches as hath been used."[19] The unit of 231 in^3 is exactly our present U.S. gallon; the smaller unit of 224 in^3 is a separate standard known to historians as the Guildhall gallon, because a measure of this capacity was traditionally kept at the Guildhall in London.

The Guildhall gallon will be mentioned again in Chapter 5 when we talk about avoirdupois weight. It was never a standard of the British government, but with the ascendancy of avoirdupois weight, it had become widely used in the retail wine trade of London by the late 1600s, a use that was to persist, still without official sanction, into the beginning of the 19th century. In a ruling of 1689, the Commissioners of the Excise reaffirmed the primacy of the 231 in^3 wine gallon as the standard for all liquids except ale and beer, which had their own standard, the ale gallon (about 282 in^3).

The ancient wine gallon was again brought into question in 1700, however,[20] and in order to prevent further confusion, the volumetric definition of the wine gallon was finally given the force of law in the act 6 Anne c. 27 §22 (1706), which decreed that

> any round Vessel (commonly called a Cylinder) having an even Bottom, and being seven Inches Diameter throughout, and six Inches deep from the Top of the Inside to the Bottom, or any Vessel containing two hundred thirty-one cubical Inches, and no more, shall be deemed and taken to be a lawful Wine Gallon; and it is hereby declared, That two hundred fifty-two Gallons, consisting each of two hundred thirty-one cubical Inches, shall be deemed a Ton of Wine, and that one hundred twenty-six such Gallons shall be deemed a Butt or Pipe of Wine, and that sixty-three such Gallons shall be deemed an Hogshead of Wine.

The engraving in Figure 3.1[21] at the beginning of this chapter shows the Exchequer standard wine gallon made according to this definition.

History of the wine gallon in England

The characteristic 63–126–252 wine cask system (Assize of London) provides a way to trace the use of the wine gallon with fair certainty about 600 years into the past.

The act 6 Anne c. 27 (1706) was the first statute to give the volume of the gallon as 231 cubic inches, but the letter from the Commissioners of the Excise to the Treasury (1688) referring to the gallon of "231 cubic inches as hath been used" makes it clear that this definition was very old by that time, so old, in fact, that its origin had been forgotten.

Its equivalence to a certain number of cubic inches aside, and also leaving aside the independent confirmation obtained from weights of wheat, the absolute size of the gallon can be tracked several centuries further back in the record by tracing the distinctive 252-gallon wine tun series with which it has always been associated. The recital of cask measures based on this tun that appears at the end of 6 Anne c. 27 simply rehearses a list that goes back, with minor variations, to the statute 2 Henry 6 c. 14 (1423) mentioned earlier (page 44).[22] Titled *The several measures of vessels of wine, eels, herrings, and salmons,* the statute (translated from Norman French) begins:

> Item, whereas in old time it was ordained and lawfully used, that tuns, pipes, tertians, hogsheads of Gascoigne wine, barrels of herring and of eels, and butts of salmon, coming by way of merchandise into this land out of strange countries, and also made in the same land, should be of certain measure; that is to say, the tun of wine twelve score and twelve gallons, the pipe of six score and six gallons, the tertian four score and four gallons, the hogshead 63 gallons, the barrel of herring and of eels 30 gallons fully packed, the butt of salmon four score and four gallons fully packed; nevertheless, by device and subtlety now late such vessels have been of much less measure, to the great deceit and loss of the King and of his people, whereof special remedy was prayed in the parliament: ...[23]

While this act of 1423 is the oldest surviving statute definition of the peculiar 252–126–63 sequence of the cask measures, the phrase "whereas in old time it was ordained and lawfully used" shows that these units were defined long before by earlier English laws now lost to us.

The statute of 1423 went on to mandate that no one could import or make a tun of wine, except it contain of the English measure twelve score and twelve gallons, the pipe six score and six gallons, and so after the rate the tertian and the hogshead of Gascoigne wine...[24]

and likewise for the barrel of herring or eels and the butt of salmon. It is noteworthy that no further definition of the word "gallon" was considered necessary; the standard on which these commercial units were based is specified simply as "of the English measure" *(del mesure dEngleterre).*

Two other statutes from the same century emphasize the age of these measures. The act 18 Henry 6 c. 17 (1439) mandated the gauging of vessels for oil and honey as well as wine "within the Realm of England," and it specified for this official function the same units of the 252-tun series just cited as "ought and were wont according to the ancient assise of the same Realm" *(doient & soloient solonc launcien assise de mesme le Roialme).* And in 1483, abuses in the measurement of a kind of wine called malmsey resulted in the statute 1 Richard 3 c. 13, which commanded

> That no manner merchant or other person whatsoever he be ... shall bring or cause to be brought into the realm, any butt of malmsey to be sold, unless it do contain in measure at the least the said old measure of 126 gallons, nor no vessels with any manner wines, whoever they be, or of what country they be, nor no manner of vessels of oil, unless the same vessels of wine or oil do contain and hold the measure and assise following, that is to say, every tun to contain twelve score and twelve gallons, and every pipe to contain six score and six gallons, every tertian to contain fourscore and four gallons, and every hogshead to contain sixty three gallons, and every barrel to contain thirty one gallons and an half, and every rundlet to contain eighteen gallons and an half, according to the ancient assise and measure of the same vessels used in this realm.[25]

Phrases such as "the ancient assise and measure" *(auncien assise & mesure)* and "the old measure" *(la veil mesure)* indicate that both system and standard were hundreds of years old by the time these laws

Chapter 3. The wine gallon

were written. The statute 7 Henry 7 c. 7 (1491) repeated the definition of a butt of malmsey as 126 gallons, and in 28 Henry 8 c. 14 §5 (1536), the entire series of wine measures implied by the act of 1483 is spelled out for the benefit of a later generation:

> And where as in the Parliament holden at Westmynster in the first yere of the reign of Kyng Richarde the thirde, among other thinges it was establisshed and ordeyned and enacted [see above] that every tonne of Wyne shulde conteign CC lij galons, every butte of Malmesey shulde conteign C xxvj galons, every pype C xxvj galons, every tercyan or poncheon lxxxiiij galons, every hoggeshed lxiij galons, every teers xlj galons [sic], and every barrell xxxj galons and di. [$31\frac{1}{2}$ gallons] and every Rondlett xviij galons and di. And that no Vessell shulde be putt to sale till it were gauged ...

In an act of a few years earlier, 22 Edward 4 c. 2 (1482), the related 84–42–21-gallon sequence was specified for casks of salmon. This statute, which set forth procedures for packing and gauging all sorts of fish, made it illegal for any person to sell any salmon

> by butt, barrel, half barrel, or any other vessel, before it be seen, except the same butt do hold and contain fourscore and four gallons, the barrel two and forty gallons, the half barrel one and twenty gallons, well and truly packed ...[26]

The same statute forbade sale of any eels

> by barrel, half barrel, or firkin, except the same barrel contain two and forty gallons, the half barrel and firkin after the same rate ...[27]

These measures for salmon and eels were reaffirmed in 11 Henry 7 c. 23 (1494), restating the above directly in English forbidding that anyone

> shuld sell nor put to sale any samon by butte barell half barell or any other vessell afor it shuld be seen, but if the same butte shuld holde and conteygne iiij^{xx} iiij [84] galons, the barell xlij^{ti} [42] galons, the di [half] barell xxj^{ti} [21] galons, well and truely packed, upon payne of forfeiture for every butte barell and di barell so lacking ther seid mesure vj s. viij d [6 shillings and 8 pence] ...

Statutes mentioning these same old cask measures go back even further than their earliest legal specification in terms of the gallon. In 27 Edward 3 stat. 1 c. 8 (1353), official gauging is mandated for "all wines red and white," and penalties are prescribed "in case that less be found in the tun or pipe than ought to be of right after the assise of the tun" *(en cas qe meinz soit trove en le tonel ou pipe qe ne deust estre de droit solonc lassise du tonel)*. Here again the phrase "of right after the assise of the tun" shows that these terms were too well understood, even in 1353, to need further clarification. The Saxon terms "pipe" and "tonel" (ton) in the Norman French original show that these names go back to a time prior to the Norman invasion.

The statute 4 Richard 2 c. 1 (1380) extended the same mandate to the gauging of sweet wines, Rhine wines, vinegar, oil, honey, and "all other liquors gaugeable."

European gallon cognates

The old English statutes cited above demonstrate an unbroken maintenance of the wine cask standards to a time well before their first mention in the act 27 Edward 3 stat. 1 c. 8 (1353), and this, in turn, proves the continuous maintenance for more than 600 years of the standard wine gallon on the basis of which all these cask measures were defined. But our gallon standard goes back much farther than that, as proved by the survival into the 19th century of what is clearly the same unit, or a very simple multiple of it (usually 2, 10, or 20 gallons), in the traditional systems of northern Europe and elsewhere.

In France, for example, a country with which the English have long had important cultural and trade connections, there was a wine unit called the velte, which directly corresponded to 2 English wine gallons (the "wine peck" implied by the 1496 statute 12 Henry 7 c. 5). This old unit retained so much vitality in the face of French metrification that it survived clear into the 1920s in commercial reference handbooks as a subdivision of the four old regional barrique or wine barrel standards then still in occasional use.[28] These were:

Bordeaux: barrique of 30 veltes
= 226.20 liters, therefore
velte = 2 gallons of 230.060 in^3

Frontignac and Sauterne: barrique of 30 veltes
= 228.00 liters, therefore
velte = 2 gallons of 231.890 in^3

Macon: barrique of 28 veltes
= 213.07 liters, therefore
velte = 2 gallons of 232.184 in^3

Bourgogne, Champagne, and Cognac: barrique of 27 veltes
= 205.46 liters, therefore
velte = 2 gallons of 232.184 in^3

Averaging these four slightly different definitions gives an implied gallon of 231.580 in^3, just one quarter of one per cent larger than our own gallon. To make the parallel complete, each velte was traditionally divided into 8 pintes, the pinte here corresponding to the unit we call a quart.[29]

The differences between these local versions of the same standard are just enough to show that the velte was not some recent adaptation to the British gallon defined as 231 in^3 but rather a descendant of an ancient parent common to both the English and the French wine units. Measurement, like language, does change when two formerly shared traditions are allowed to evolve separately, but, even more so than language, a culture's metrology changes very slowly. The differences between the four regional velte standards and the English wine measure show that the common ancestor of all these units must have been very old.

Besides France, the place one would expect to find a unit related to the old wine gallon would be Germany, in particular Saxony. The system that continued to be used in Saxony until the establishment of the metric system did not, as far as I know, include the gallon itself, but instead was based on a unit called the anker that corresponded to 20 of our gallons. The anker that survived into the 19th century in Saxony was defined as 54 gauger's kanne of 1.404 liters apiece,[30] which means that the anker was 20.0285 of our gallons or 20 gallons of 231.329 in^3; compare this with the 231.580 in^3 gallon implied by the early 20th-century value of the average French velte.

Other units corresponding to 10, 20, or 40 of our gallons were in use among the countries that traded along routes in the North Sea and the Baltic Sea. In Denmark and Norway, the main official unit of commercial trade up through the end of the 19th century was called the anker, as in Saxony, but this anker was about half as large as the Saxon unit, corresponding to 10 gallons of 229.930 in^3.[31] Until the 1930s, the unit of volume in the kingdom of Estonia was the vaat of 150.664 liters, which is 40 gallons of 229.852 in^3, and in what was formerly the neighboring Baltic state of Livonia, the vaat was $\frac{1}{6}$ of the oksaam of 229.580 liters, making the vaat 10 gallons of 233.497 in^3.[32] These three examples point to an average Baltic/North Sea standard of 231.093 in^3, virtually identical to our wine gallon.

Under the name anker—just as in the Germanic countries—the measure of 10 gallons remained in some American arithmetic books right through the 19th century, and of course it continues to exist down to the present as a very common unit for industrial liquids in the U.S. The larger significance of this unit of 10 gallons and its multiples will become apparent further on when we examine the relationship between the gallon and the length standard embedded in our old system of land measurement.

If you include our own gallon, the nine survivals just cited average 231.325 in^3 with a standard deviation of 1.17 in^3 or 0.5 percent. Taken together, these obvious family associations lead to the conclusion that an ancestor of our gallon—or more strictly speaking, an ancestor of our commercial unit of 10 gallons—was for many centuries one of the common volume standards of northern Europe.[33] And small as they are, the differences between the different 19th-century national standards show that these siblings split off from one another many centuries ago—certainly no more recently than late Roman times—and that the common ancestor must be extremely old.

Fortunately, we don't have to rely on logic to establish the existence of the gallon in Britain during the Roman occupation; we are lucky enough to have physical evidence.

Chapter 3. The wine gallon

The Carvoran measure

In 1915, one of the two best examples of ancient Roman capacity measures still in existence was found in the marsh outside the Roman-British fort at Carvoran Magna on Hadrian's Wall, which was for several centuries the northern boundary of Roman occupation in Britain. The artifact is a grain measure, crafted with typical Roman expertise out of bronze about a quarter of an inch thick. It is very well preserved (as objects can be in northern marshes) and, unlike many ancient capacity measures, it is clearly inscribed in a way that for the most part leaves no doubt as to its interpretation.[34]

In form the Carvoran measure is a truncated cone, about the same size and shape as an ordinary galvanized pail but with the large end at the bottom, probably to provide greater stability as the grain was being leveled (Figure 3.4, page 51). To ensure even leveling, the measure is equipped with a three-prong leveling bar that spans the open top, providing a reference plane across the opening and preventing variations in apparent capacity caused by the application of too much pressure on the contents during the leveling operation.[35]

The bottom of the measure is a flat disk, made as a separate casting and recessed into the base of the measure so that it never actually touches the ground. This bottom is held in place by four symmetrical prongs that project inward about an inch from the sides of the measure and support the bottom from below, the bottom being secured to each prong with a businesslike touch of solder. A bronze rod, brazed in place at each end, spans the vertical distance between the leveling plate at the top and the inside center of the base plate, wedging the bottom in place and preventing the leveling plate at the top from possibly becoming bent inward in hard use. Although the bottom is not soldered or sealed to the sides in any way—this is, after all, a grain measure—the pieces fit together so closely that the vessel is virtually watertight.

Deeply incised on the measure, in Roman capitals about $\frac{3}{4}$ inch in height, is the following legend:

IMP [erasure] CAESARE
AUG . GERMANICO . XV . COS
EXACTVS . AD . S . XVIIS
HABET . P . XXXIIX

That is,

IMPERATORE [erasure] CAESARE
AUGUSTO GERMANICO XV CONSULE
EXACTUS AD SEXTARIOS XVII SEMIS
HABET PONDO XXXVIII

which translates as

EMPEROR [name erased] CAESAR
AUGUSTUS GERMANICUS IN HIS
FIFTEENTH CONSULAR PERIOD
EXACTLY $17\frac{1}{2}$ SEXTARII
IT WEIGHS 38

Although the personal name of the emperor was sanded or filed off the measure some time after it was made, the remaining titles (according to Skinner) fit no other emperor but Domitian. Never a promising ruler, Domitian eventually became paranoidally insane and behaved so viciously during the last three years of his life that after his death (reportedly arranged in self-defense by his wife), his name was erased almost everywhere. The reference to Domitian's fifteenth consular period dates the manufacture of the Carvoran measure to the period A.D. 90–92.

The enigmatic "it weighs 38" of the last line of the inscription could refer to any of a number of roughly pound-sized units, depending on whether the weight referred to is the weight of the measure itself, the weight of the measure's contents (of some substance or other), or the weight of the measure with contents included. Most likely it refers to the weight of the vessel itself (what we would call the tare weight), but in the absence of a handle, now missing, that was formerly attached to the measure, it is impossible to know its original weight and therefore to draw any firm conclusion about this.

No such ambiguity attaches to the volumetric part of the inscription, however. The sextarius was a common Roman unit of both liquid and dry measure, and its embodiment in so well-wrought a vessel constitutes the best remaining evidence of the volume of the sextarius used in Roman trade with its northern provinces.

The capacity of the Carvoran sextarius was established with great precision in a series of determinations carried out by the British standards department in 1948. First, the measure was gauged by measuring the volume of its contents in rape seed, a

Chapter 3. The wine gallon

Figure 3.4: The Carvoran Measure

common technique for finding the capacity of measures and storage vessels. It was found to have a capacity equivalent to the volume of 174,102 troy grains of water. Then water was used, and filled to the brim the measure was found to actually contain (under standard conditions) 174,597 troy grains of water. In other words, it was found that a slightly greater volume of water will fit in the measure than seed, presumably because of the ability of water molecules to occupy microscopic pits or indentations on the inner surface of the measure.

It is an even argument as to which method is best for judging the capacity of a grain measure, but it is safe to say that the intended volume of the Carvoran measure must be somewhere in the narrow range (less than one third of a percent difference) delimited by these two gaugings. Dividing by the factor of $17\frac{1}{2}$ inscribed on the measure and then converting to volume using the British definition of standard water at 62 °F,[36] the standard Roman trade sextarius defined by the Carvoran measure is found to have a volume of 39.4342 to 39.5463 in^3, depending on whether the dry or the wet method is used for gauging the measure.

The trade sextarius is thus proved to be distinct from the sextarius used in Rome, which is known to have been about 13 percent smaller.[37]

Looking at the Carvoran measure in terms of Roman standards shows how carefully it was made. It's immediately evident that the trade sextarius embodied in the Carvoran measure was derived from the cubic Roman foot, no doubt for ease in calculating the capacity of cargo spaces on ships. The relationship is simply

40 trade sextarii = 1 cubic Roman foot

Modern estimates of the mean Roman foot range from a low of 11.64 inches to a high of 11.65 inches. By comparison, the foot implied by the seed volume of the Carvoran sextarius is 11.6407 inches, and by the water volume 11.6517 inches.

Half of a sextarius was called a hemina, and it follows from this relationship that there were 80 trade heminae in a cubic Roman foot. This is particularly significant in light of the widely attested definition of the Roman libra (pound) as $\frac{1}{80}$ the weight of a cubic Roman foot of water.[38] In other words, the Ro-

Chapter 3. The wine gallon

man trade system embodied in the Carvoran measure was based on the relationship

1 trade hemina of water weighs 1 libra

It turns out that the weight of a trade hemina full of water does, in fact, correspond quite accurately to a well-established Roman weight standard. If we split the difference between the two gaugings of the Carvoran measure and use their average to represent the narrow range of possible capacities, the Carvoran hemina ($\frac{1}{35}$ of the measure) has a volume of 19.7451 in^3 (giving a corresponding Roman foot of 11.6462 inches), and this volume of water under the conditions used to define the Imperial gallon weighs 4981.41 grains or 322.790 grams. Since the Roman pound consisted of 12 ounces (unciae), and each ounce was divided into 24 scruples (scripula), the weight standard defined by the Roman foot, as embodied in the hemina of the Carvoran measure, is

pound (libra)
  = 4981.41 grains (322.790 gm)
ounce (uncia)
  = 415.118 grains (26.8992 gm)
scruple (scripulum)
  = 17.2966 grains (1.12080 gm)

This implied Roman trade standard is in accurate agreement with one of the best of the few Roman weight sets still in existence, a beautifully made series of multiples of the scripulum from Lyons in Roman Gaul (modern France). The weights agree on a mean scripulum of 17.283 ± 0.01 troy grains (1.11992 ± 0.0006 gm), which differs from the scripulum arrived at above by less than one tenth of one percent.[39] And the weight units implied by the Roman trade measures are further attested by 18th- and 19th-century survivals of these units or their simple multiples all over Northern Europe, Russia, India, and North Africa, i.e., in all the directions in which Roman trade extended.

The point of this excursus (aside from demonstrating the accuracy of the ancient standards underlying the network of metrological relationships to which our Customary System is heir, and the care with which they were maintained) is to establish that the Carvoran measure is what it appears to be: an official product of the most technologically advanced society of the ancient world at the height of its powers. Together with the system of weights and measures it implies, the Carvoran artifact is just the sort of work we would expect from a nation anciently reknowned for its bronze work and later noted for its civil engineering.

But all this just throws into sharper relief the real strangeness of the capacity inscribed so prominently on the vessel. Why $17\frac{1}{2}$ sextarii? The number $17\frac{1}{2}$ has no more significance in the Roman system than it does in ours, and the Carvoran measure is just as puzzling as would be a measure inscribed

U.S. BUREAU OF STANDARDS
EXACTLY $17\frac{1}{2}$ QUARTS

The stated capacity of the Carvoran measure is just enough bigger than the usual Roman measure of dry capacity, the modius of 16 sextarii, to make it clear that its makers had in mind something completely different.

The native gallon

The mystery of the Carvoran measure is solved when we consider the way the Romans dealt with the indigenous systems of weight and measure in their newly conquered territories. Their habits are known from an instance involving the Northern foot, which you learned about in Chapter 1. When the Roman general Drusus encountered the Northern foot (13.2 inches) among the Germanic tribes north of Italy, he did not order its replacement by the Roman foot of 11.64–11.65 inches; on the contrary, the Northern foot was officially adopted in 12 B.C. as the standard for the northern settlements. But this "Drusian foot" was defined for Roman purposes as 18 Roman digits (that is, 13.10–13.11 inches instead of the actual 13.2), the digit being $\frac{1}{16}$ of the Roman foot.[40]

In other words, the native tribes were required to accept the nearest whole-number Roman approximation, executed precisely in Roman units, as a reasonable substitute for their own when calculating land holdings for tax purposes. And I believe that the Carvoran measure shows that this is what happened with the native units of capacity that the

Romans found when they conquered Britain. The Carvoran measure is simply 3 native wine gallons, the gallon being approximated at a value of 230.033 to 230.687 in^3 (depending on how the measure is gauged). This is a difference from the actual wine gallon, as it has come down to us, of less than one half of one percent, which is not bad considering that its makers never intended to reproduce the native unit of 3 gallons exactly but to realize its closest approximation in whole numbers of precisely defined Roman units (35 trade heminae)—which they did, as we have seen, with formidable accuracy.

The only remaining mystery is why the Romans chose a measure of 3 gallons rather than something more consonant with the traditional binary structure of the gallon system. The simple answer is that a measure of 2 or 4 gallons would not have worked out as accurately in whole numbers of heminae, which the makers of the Carvoran standard seem to have been determined to accomplish. But there are other levels to the meaning of this 3-gallon unit. As we've already seen, the gallon system was thoroughly binary but not exclusively binary; the tradition allowed supplementary units, multiples of the gallon that included factors of 5 or 10 (as in the various anker-size units) and sometimes even 3, as in the tierce and the tertian. At a deeper level still, the 3-gallon unit turns out to have a place in our Customary system of weights, as will be seen in Chapter 5.

Before the Romans

The Carvoran measure points back to a time far earlier than the date of the measure itself. It's very unlikely that the Romans would have chosen to work in a unit as unwieldy as $17\frac{1}{2}$ sextarii unless the native standard was already too firmly embedded in the local culture to change, and this indicates that the wine gallon standard is much older than the Roman occupation.

In fact, as we shall see, the wine gallon is convincingly related to not one, but at least three ancient standards of volume. Instead of thinking of the gallon as the descendant of a single tradition, it's probably best to think of it as expressed or implied by an interconnected complex of ancient standards supported by various trade relationships, just as in the case of the weight standards we looked at in Chapter 2.

Years of study have convinced me that all the major systems of ancient weights and measures bear very simple whole-number relationships to each other, at least in measures of weight and capacity.

The first thing one discovers upon investigating ancient mathematical abilities is that, with the exception of some Sumerian mathematicians and Babylonian astronomers aided by a positional base-60 system of numeration, the ancients had a hard time dealing with fractions. For example, papyri show that the Egyptians had to calculate using fractions of all kinds but (with the exception of $\frac{2}{3}$) could only represent them as a series of unit fractions, that is, fractions with 1 for the numerator. So where we would write $\frac{2}{73}$, say, an Egyptian would have to write the equivalent of $\frac{1}{60} + \frac{1}{219} + \frac{1}{292} + \frac{1}{365}$.[41] A demotic papyrus of arithmetic exercises with the numbers we would write as $\frac{2}{5}$ and $\frac{45}{63}$ shows them being added as

$$(\tfrac{2}{3} + \tfrac{1}{21}) + (\tfrac{1}{3} + \tfrac{1}{15}) = (1 + \tfrac{1}{10} + \tfrac{1}{70})$$

and subtracted as

$$(\tfrac{2}{3} + \tfrac{1}{21}) - (\tfrac{1}{3} + \tfrac{1}{15}) = (\tfrac{1}{4} + \tfrac{1}{28} + \tfrac{1}{35}).[42]$$

The difficulty this introduced into calculations can scarcely be imagined.[43]

For this reason, any two ancient trading cultures would of necessity begin to adapt their measures to each other, not to abandon the original units, but to nudge them into simple whole-number ratios between the two systems in all contexts where they needed to be directly compared or converted—when a trader from a distant place began to sell in one sort of local units what he had previously bought in another, for instance, or a money-changer went to sell gold objects made in units of one system to people who used another. Since units of length were employed primarily in constructing buildings and measuring land, they did not undergo the same process of direct adaptation seen in weights and capacity measures, but instead formed relationships indirectly through measures of volume as traders used lengths to calculate the capacities of ships and granaries.

Chapter 3. The wine gallon

Several important length standards of the ancient world demonstrate this kind of interrelationship. The lengths themselves appear to be unrelated, but the volumes they create tell a different story. Consider, to take the most striking example, the Egyptian cubit and the Sumerian foot. These were probably the most important length standards of the ancient world, and they appear on the surface to have nothing to do with each other.

In his masterful article in the 11th Edition of the *Encyclopædia Britannica,* Petrie put the Egyptian cubit at 20.63 inches,[44] or "most accurately 20.620 in the Great Pyramid."[45] Not long after, the Assyriologist Thureau-Dangin established through the study of ancient bricks the value of the foot and cubit in bronze-age Sumeria and later Babylonia to the nearest millimeter. These measurements put the Sumerian foot at its generally accepted value of 330 mm (12.99 inches) and the Sumerian cubit of $1\frac{1}{2}$ feet at 495 mm (19.49 inches).[46]

You will search in vain for any simple relationship between the Egyptian cubit of 20.62 or 20.63 inches and the Sumerian foot of 12.99 inches; but cube them and you will instantly find that the Egyptian cubic cubit is 4 Sumerian cubic feet, exactly. That is, if the Sumerian cubic foot is the cube of 330 mm, then 4 Sumerian cubic feet correspond to the cube of an Egyptian cubit of 20.624 inches.[47]

It's important to understand that the ancients didn't necessarily ever think of the relationship between these two units directly, because there would very seldom have been an occasion to compare length units from different cultures. Rather, the relationship is a *logical consequence* of the associated weight and capacity equivalents on either end of a trading relationship.

Another example is the relationship of the Northern foot to the ancient Greek foot. The Northern foot, as you saw in Chapter 1, is historically about 13.22 inches and persists in our land measures at 13.2 inches. The Greek foot was widely distributed in ancient times and continued right up to the early 20th century as the official foot of the Austro-Hungarian empire at a value of 12.444 inches.[48] A little work with a calculator shows that 5 cubic Northern feet equals 6 cubic Greek feet. Ancient weight standards show similar simple relationships.

It's not a surprise, then, that our ancient wine gallon can be clearly related to more than one ancient volume standard. Two units of our traditional system that most clearly exhibit this kind of relationship to ancient ancestors are the unit of 10 gallons and the unit of 32 gallons; each corresponds directly to the cube of a major ancient measure of length.

Back to the land: the 10-gallon anker

We've already seen the wide distribution of the anker or 10-gallon unit in Northern European traditions as well as in the U.S.

This unit is quite simply a cubic Northern foot, the basis of all our land measures. If the gallon is 231 in^3, then the corresponding Northern foot is 13.219 inches.[49]

It follows from the definition of the wine gallon that a cubic Northern foot of wheat weighs 80 troy pounds, echoing the Roman tradition noted earlier of a cubic Roman foot of water weighing 80 Roman pounds. And here is the connection between our land measures and the measures we devised for the products of the land:

1 cubic Northern foot contains 10 wine gallons or a quantity of wheat weighing 80 troy pounds

And because the Northern foot and the Greek foot are related by cubic measure:

100 wine gallons (10 cubic Northern feet) is the same as a dozen cubic Greek feet

It appears to me that the association between the gallon and the Northern foot is not confined to England. In Chapter 1, you saw Chinese land measures almost identical to our Customary system; in Indochina, we find an old measure that nicely demonstrates the relationships between the Northern foot and the wine gallon. There, until finally replaced by the metric system in 1903, the primary measure of capacity was the gia, defined as half of a cubic thuoc moc. The thuoc moc was 424mm, which makes the gia 38.112512 liters.[50] This is equivalent to 10 gallons of 232.58 in^3; that is, to put it another way, the gia was a cubic Northern foot, the Northern foot being 13.25 inches. Another unit with what appears to be the same basis is the Siamese (Thai) thangsat, defined by law (ca. 1875)[51] as 2000 cubic nin or 1157.4074 in^3; this is 5 gallons of 231.48148 in^3 or half of a cube with sides measuring 13.23 inches.

The Ur-barrel of 32 gallons

The appearance of the wine gallon in the Carvoran measure and its clear relationship to both the Northern foot and the troy pound strongly suggests that it is probably as old as these other two standards. In fact, I believe that we can trace its history back to the late bronze age by looking again at its characteristic binary mode of division.

Just as in the case of weights, binary division and composition are the most natural and primitive way to create a series of capacity measures. Before measurement as we ordinarily understand it—before counting, even—comes direct comparison of two quantities. If we arbitrarily begin with a certain container and call that a unit, it is trivially easy to duplicate its capacity, and then equally simple to form a binary step upward in the series by creating a vessel that holds these two equal units combined, and so on.

You have already seen the strong binary aspect of our liquid gallon system going downward and of the cask system going upward. In Chapter 4 you will see the dry gallon itself (a different standard) form a binary series upward to form the bushel and other measures for dry substances.

From old English statutes, it's evident that a binary cask system based on the wine gallon and exemplified by a 32-gallon barrel existed alongside the wine cask system based on a barrel of $31\frac{1}{2}$ gallons discussed earlier.

The 32-gallon barrel, used primarily for packing meat and fish, makes its first statute appearance in 22 Edward 4 c. 2 (1482), which mandated (the original is in Norman French)

> that no merchants nor other person set any herring to sale by barrel, half barrel, or firkin, except the same barrel contain 32 gallons, the half barrel and firkin after the same rate . . .[52]

This statute was reaffirmed in 11 Henry 7 c. 23 (1494).

The earlier (and uncharacteristic) reference to a herring barrel of 30 gallons notwithstanding—see 2 Henry 6 c. 14 (1423) quoted above (page 47)—the role of the 32-gallon barrel as the traditional unit for packing herring is made clear by a portion of a later statute, 13 Elizabeth 1 c. 11 §4 (1570). This law was apparently passed to resolve a conflict resulting from a confusion of the wine gallon (231 in^3) with the substantially larger Winchester gallon, which we will look at in Chapter 4 in connection with the U.S. Customary System for dry measures. In the words of the statute:

> And where yor Subjectes using the Trade of fyshing for Hearring have of many yeres, and tyme out of mynde used to packe theyr Hearring in Caske of Barrels conteyning aboute thyrtey two Gallons of usuall Wyne Measure, and with such Assise hath ben usually gaged and allowed at your honorable Cytye of London, and do conteyne the said Measure of xxxij Gallons, according to suche usuall Brasse Measure as is out of your honourable Court of Exchequer delyvered to your said honorable citie of London; wch Measure yet hath lately ben quarrelled at by certayne Infourmers, for that the same conteyne not two and thyrtey Gallons by the old Measure of Standerd [i.e., the Winchester standard, Chapter 4], wch they never did, though peradventure the extremitie of old Statutes in Wordes by some Mens construction might be stretched to requyre so muche; And for that the usual Barrels nowe be as great as ever wthin the tyme of any memory they have ben knowne to be, and the Alteration thereof woulde be a great Decay and Perill of undoing to the saide Fysshermen; It maye also please your most Excellent Matie that it be also ennacted and declared, That the sayde Assise of thyrtey two Gallons of Wyne measure, wch is about eight and twentey Gallons by olde [Winchester] Standerd well packed, and conteyning in every Barrel usually a thousande full Hearringes at the leaste, is and shalbe taken for

Chapter 3. The wine gallon

good true and lawfull Assise of Hearring Barrels....

The 32-gallon barrel was assumed as the standard unit of measure for codfish, ling, hake, salt beef, and salt pork, as well as all kinds of herring, in laws as late as 38 George 3 c. 89 (1797) . Considering the importance of fish to England's economy and the role of salted meat in the diet of that time, these laws make it clear that the barrel of 32 wine gallons was an important unit.

The 32-gallon barrel was specified for another significant commercial substance, honey, by the statute 23 Elizabeth 1 c. 8 §4 (1581), which made it illegal for any person to

> fill and sell or cause to bee filled or soulde or offered to be solde, any Barrell Kilderkine or Fyrkyn with Honny, for or in the Name of a Barrell Kilderkine or Fyrkine, conteyning lesse then Thirtye two Wyne Gallons the Barrell, Syxtene Wyne Gallons the Kylderkine, and Eyght Wyne Gallons the Fyrkine....

The barrel of 32 gallons was also mandated for the sale of soap by the act 23 Henry 8 c. 4 §6 (1531).

Holinshed, writing in 1577, says that "our English measures" in the larger sizes were gallon, firkin of 8 gallons, kilderkin of 16 gallons, and barrel of 32 gallons, the series based on the barrel of $31\frac{1}{2}$ gallons being used for foreign (wine) trade.[53]

So strong was the barrel of 32 wine gallons in the English tradition that it continued to exist, disguised in Imperial units, well into the 20th century. A 1913 act of Parliament defined the herring barrel as $26\frac{2}{3}$ Imperial gallons,[54] which is simply the 32-gallon barrel restated in Imperial units.[55]

In the American colonies, the barrel of 32 wine gallons turns up in a 1746 South Carolina statute as the barrel for tar,[56] a 1768 Georgia statute as the barrel for beef and pork,[57] a 1784 North Carolina statute as the barrel for pitch, tar, and turpentine (naval supplies important to the colony's economy at that time),[58] and an 1819 Alabama law as the barrel for tar and turpentine.[59]

Though we no longer think of it in connection with herrings or honey, the 32-gallon barrel maintains a lively existence on the streets of American cities: it is still a popular size for trash barrels, in particular, the ones used for recycling. So the barrel of 32 wine gallons continues to be part of our everyday experience (Figure 3.5, page 57).

The thoroughly binary nature of the main gallon sequence creates a strong presumption that the oldest system of units for casks was this one based on a barrel of 32 wine gallons, a presumption made all the stronger by its specification, without further explanation, in the statute of 1482 and by the statement in 1570 that this barrel had been in use "of many years, and time out of mind"—i.e., from time immemorial—and that such barrels were then "as great as ever within the time of any memory they [had] been known to be." And the binary sequence upward from the gallon parallels the binary sequence upward for the dry measures we'll be looking at in Chapter 4, which argues strongly that the binary sequence is the native system.

Thus, there is every reason to believe that unlike the barrel of $31\frac{1}{2}$ gallons, which appears to be a medieval innovation, the barrel of 32 gallons is as old as the gallon itself, which we know from the Carvoran measure to have been well established in Britain by the end of the first century A.D.

If we now follow this line of reasoning one step further and ask whether there is a known, widely used, pre-Roman standard of capacity corresponding to the 32-gallon barrel, the answer is an immediate *yes*. The barrel of 32 wine gallons (U.S. gallons) is identical to what is apparently the oldest, and certainly one of the best established, units of volume in the ancient world, the Sumerian cubic cubit. This standard is named as the unit of measurement in hundreds of clay tablets containing calculations dealing with large volumes.

As noted earlier, the generally accepted value for the Sumerian cubit is 495 mm. By contrast, a barrel of 32 U.S. gallons under our present definitions has the volume of a cube with sides measuring 494.790 mm. To put it another way, if we take the archaeologically determined figure of 495 mm for the Sumerian cubit as exact, then the gallon formed by binary division of the cubic cubit would contain 231.294 in^3.

This correspondence goes beyond a coincidental equality of two isolated units; the binary series of our customary units of capacity fits a binary division of the cubic cubit all the way up and down. Our

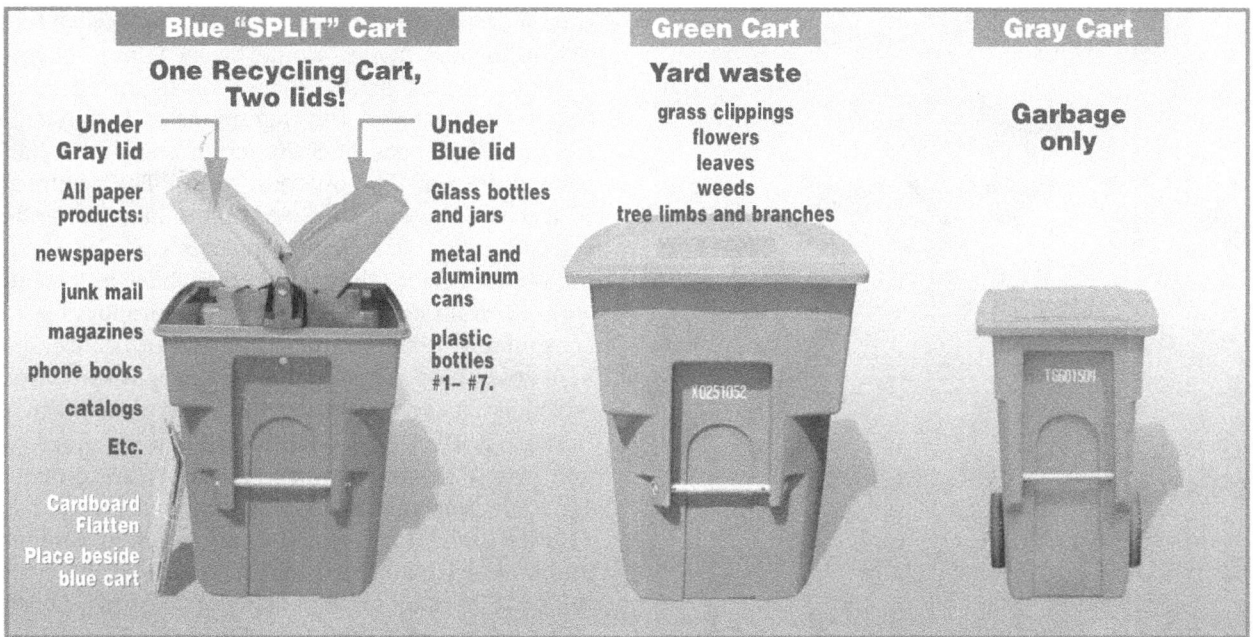

Figure 3.5: Recycling carts (from the web site of the City of Pleasanton, California). These containers of 32, 64, and 96 gallons provide a nice illustration of the way that standards can continue in existence even when all knowledge of their origin has long been forgotten

ordinary measuring cup is the volume of a cube $\frac{1}{8}$ cubit on a side; the half-gallon, a cube $\frac{1}{4}$ cubit on a side; 4 gallons, a cube $\frac{1}{2}$ cubit on a side; the 32-gallon barrel a cubic cubit; and the old tun of this standard, 256 gallons, a cube with sides equal to a Sumerian double-cubit of 3 Sumerian feet.

The relationship of the wine gallon system to Mesopotamian originals goes beyond its correspondence with the Sumerian cubic cubit, however; we have direct evidence of Babylonian versions of the wine pint and the wine quart themselves.

Babylonian pints and quarts

The Old Babylonian version of our wine pint is exemplified by the famous vase of King Entemena of Lagash in the Louvre, ca. 2400 BC (Figure 3.6, page 58). This elaborately decorated silver vessel has a capacity, filled to the top, of 471 mL,[60] which is 28.74 in^3—the pint of a wine gallon of 230 in^3.

A Neo-Babylonian bottle from Persepolis, about 600 B.C., is marked with its capacity, and a careful measurement carried out at the University of Chicago in 1952 puts the unit called a qa implied by this vessel at 944.9 mL, almost exactly twice the volume of the vase of Entemena.[61] "We believe that it will hardly be possible to determine the equivalents of the ancient measures concerned more accurately," the authors say, and "we may mention parenthetically that this figure is surprisingly close to the equivalent (946.3 ml.) of one U.S. liquid quart"—not so surprisingly in the context of the whole history of this standard. The Persepolis unit is one quarter of a wine gallon of 230.6 in^3.

Whether the wine gallon standard could have come into ancient Britain directly from Mesopotamia is an open question. It is certainly not impossible; we know that there were bronze-age trading relationships that extended all the way from Britain to the Near East. For example, bronze-age craftsmen in Wessex developed a specific technique for perforating amber spacer plates in multi-strand necklaces, and these are found in only one other place in the world—Mycenaean Greece.[62] And there is ample evidence of Mesopotamian measures

Chapter 3. The wine gallon

Figure 3.6: Vase of King Entemena, ca. 2400 BC (from Wikimedia Commons)

sure or "megalith yard" of 2.720±0.003 English feet (32.64 inches), which would be $2\frac{1}{2}$ Sumerian feet of 13.06 inches or 331.6 mm (it is known from the clay tablets that the Sumerians and Babylonians did use a pace of 5 Sumerian feet and a half pace or step of $2\frac{1}{2}$ Sumerian feet). The accuracy with which this standard was maintained from one end of Britain to the other argues strongly for a central point of orgin, though whether this was located within Britain or on the Continent is impossible to determine.[65] And a large collection of bronze-age Irish gold objects shows Sumerian weight standards mixed in with several others from the middle east.[66]

There is even evidence that the cubic Sumerian foot, which would have been an important bronze-age trade unit, survived in some parts of England as a bushel standard; it was strong enough in popular tradition to have been established in a Connecticut law of 1800 at an implied value of 2198 in^3, which is the volume of a cube 13.0020 inches or 330.250 mm on a side.

The Connecticut statute provides "That the brass measures, the property of this state, kept at the Treasury, that is to say, a half bushel measure, containing one thousand and ninety-nine cubic inches, very near, a peck measure and half peck measure, when reduced to a just proportion, be the standard of the corn measures in this state...."[67] That this value for the bushel was not some kind of accident is demonstrated by the descriptions of the brass measures ordered to be constructed as state standards: a two quart standard was to measure 4×4.5×7.63 inches, a quart standard was to measure 3 inches square by 7.63 inches in height, and a pint standard was to measure 3 inches square by 3.82 inches in height. The dimensions of the two-quart and one-quart measures imply a bushel of 2197.44 in^3 (a cube with sides 13.0009 inches or 330.222 mm in length), and so would the dimensions of the pint if its height hadn't been rounded off to two places in dividing 7.63 by half.

The fact that the Connecticut bushel is a cube almost exactly 13 inches on a side doesn't lead anywhere, as there is no earthly reason to have chosen 13 inches, and if a cube 13 inches on a side had been intended, then a corresponding bushel of 2197 in^3 or half bushel of $1098\frac{1}{2}$ in^3 would have been specified. It's certainly possible that the traditional

surviving in Britain and elsewhere in Europe.

According to Petrie, who surveyed the structure in 1922,[63] the Silbury burial mound, which is the largest stone circle in England, was laid out with a foot of 13.0 inches, that is, a Sumerian foot (13 inches is 330.2 mm); the late neolithic date for Silbury (ca. 2400 B.C.) is about right for transmission from early bronze-age Sumeria. In 1877, Petrie[64] had found seven medieval English examples averaging 12.97 inches (329.4 mm) but was unable to assign this unit an origin, as at that date no one had yet discovered the Sumerian foot.

Based on examples from Stonehenge and hundreds of less famous neolithic stone circles and other features, Alexander Thom has established the existence throughout ancient Britain of a mea-

English measure from which the Connecticut bushel was presumably inherited rationalized the unit as a cube 13 inches on a side somewhere along the line; but this doesn't explain anything, it just pushes the question back in time. If the Connecticut bushel didn't come from some survival of the Sumerian cubic foot, then its origins are a complete mystery; all of the other colonies used the Winchester bushel of 2150.42 in^3 (Chapter 4) or a closely rounded-off version of this value as the standard measure for grain. The Sumerian cubic foot bushel remained law in Connecticut until 1850.

Such a survival would hardly be unique. The Sumerian foot survived at exactly its ancient value (330 mm) in Portugal till replaced by the metric system in 1864[68] and in Portuguese dependencies (Brazil, Macau) until the mid-20th century.[69] The Sumerian double cubit survived less accurately but still recognizably in Persia (Iran) and India into the mid-20th century as well.[70]

In 19th century Genoa (Italy) the cubic cubit itself survived as the unit of dry measure. By that time it was called the mina and valued at 120.70 liters,[71] which is 32 gallons of 230.174 in^3. The subunits of this measure demonstrate a binary evolution from the cubic cubit, as in the case of our gallon, but with the typical Roman 12 thrown in: $\frac{1}{8}$ mina was the quarto, corresponding to the English unit of 4 gallons (half firkin), and $\frac{1}{12}$ quarto was a gombetta, corresponding to $\frac{1}{3}$ of the wine gallon.

In Peru, the aroba for wine and spirits—doubtless a colonial preservation of an old Spanish standard—was in the 20th century still used in the interior of the country at a value of 6.70 British Imperial gallons,[72] which is 8 gallons of about 232 in^3. At the other end of the Arabic axis from Spain we find the 19th century gudda or cuddy of Arabia, which measured 7.567 liters or 1.999 U.S. gallons.[73] The cuddy was subdivided into 8 noosfia (our quart) and each noosfia into 16 vakia (each vakia equivalent to 2 of our fluid ounces). Geographical proximity makes the inheritance from Mesopotamia easy to see in this case.

Including he French wine velte noted earlier (average velte equal to 2 wine gallons of 231.580 in^3), the Italian, Peruvian (Spanish), and Arabic examples just cited point to a common Mediterranean version of the gallon standard representing the diffusion of a binary subdivision of the cubic Sumerian cubit and giving an average gallon ($\frac{1}{32}$ cubic cubit) of 231.16 in^3. Such a system makes it very easy to estimate the number of retail capacity units contained in a storehouse whose size is measured in Sumerian cubits.

While it's not impossible that the gallon came into Britain directly from Mesopotamia, however, it is more likely that it entered via a trading relationship with a Middle Eastern cultural complex that included other ancient civilizations—Egypt, for example.

The tomb of Hesy

The Egyptian version of the gallon system can be deduced from a remarkable mural painting found in the tomb of Hesy, a high Egyptian official of the third dynasty king Neterket (Sa-nekht), about 2650 B.C.;[74] the illustration from the original report is reproduced in Figure 3.7 (page 60).[75] As described by Flinders Petrie, the painting shows

> two series, each of 14, of graded measures of capacity: the upper series made of wooden staves, coopered with top, bottom, and middle bands; the lower series, coloured red, probably of thin beaten copper. On comparing these two sets they are seen to be of the same series of sizes in both materials. As the copper must have been thin, the wood must also have been very thin for the contents to be alike. The wooden set is evidently for dry measure, the metal set for liquids. Each measure nearly follows the modern rule that the depth is equal to the diameter.[76]

The close-up in Figure 3.8 (page 61) gives some idea of the care with which the measures were represented.[77] (The two objects that look like lids are actually *strikes*—the wooden bars or laths that have been used for thousands of years to level off a grain measure without compressing the grain.) As noted in the original report,

> The wood graining on some of the tubs is done with great skill and is closely copied from reality; every fibre is followed to the edge of its

Chapter 3. The wine gallon

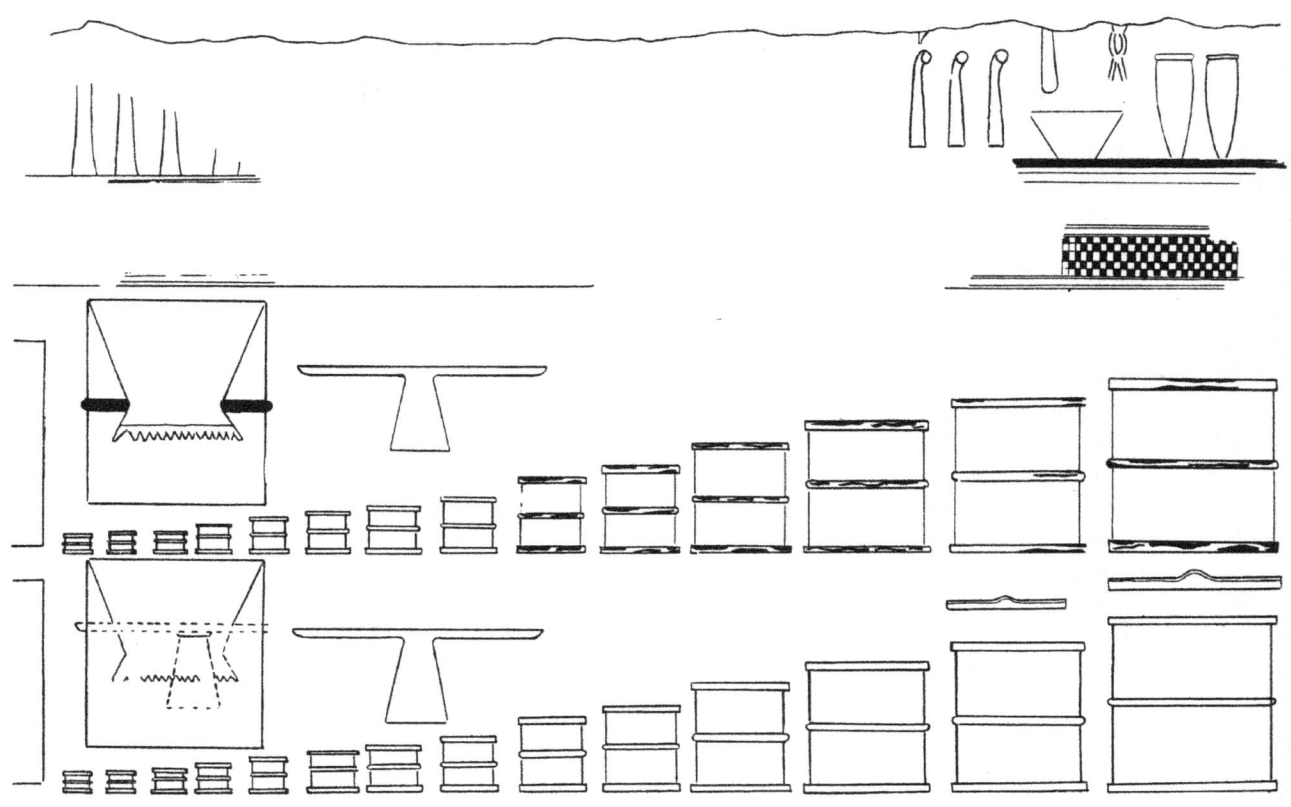

Figure 3.7: Capacity measures from the tomb of Hesy, ca. 2650 B.C.

plank and in the acute angles where the fibres turn round a knot dark shading is put in. Ths graining is best shown in the largest tub but one, where five planks are clearly distinguished.

It will be noted that real cooper's work is intended,—barrels with bevelled staves. The hoops at top and bottom are square in section, those in the middle are rounded.[78]

In the typical Egyptian fashion, the vessels are shown in an exact side view, which makes outside measurements easy, and since all the tomb paintings were drawn life size (as confirmed by depictions of weights elsewhere in the mural, calculations of the weight of which based on size and density of the materials used correspond well to actual weight data), the outside dimensions can be gauged with fair accuracy.

Petrie's remark regarding the similarities between the wooden and copper measures notwithstanding, the lack of information regarding the thickness of the wood makes it pointless to attempt to determine their exact capacity, but no such difficulty attaches to the copper measures; the expense of copper would have required the walls of these vessels to have been as thin as possible, and this is confirmed by the general run of ancient copper vessels. The care with which the wooden measures were rendered justifies us in assuming an equal level of care in the execution of the images of the copper ones as well, and their capacities can therefore be estimated with reasonable certainty.

The copper measures turn out to form two interleaved series of units, each series exhibiting a binary set of relationships. According to Petrie, one series is based on a unit of 28.8 in^3 and the other on a unit of 21.6 in^3.[79]

60

Figure 3.8: Detail of the largest measures from the tomb of Hesy

Chapter 3. The wine gallon

The unit of 28.8 in³ is our wine pint again, corresponding to a gallon of 230.4 in³. Thus, the binary series on this unit corresponds throughout with the units of our Customary gallon system, the measures of the tomb painting representing $\frac{1}{2}$ pint through 4 gallons, and it also aligns perfectly with the binary divisions of the Sumerian cubic cubit noted earlier.

The other standard displayed in the tomb mural, 21.6 in³, is that of the Phoenecian/Syrian kotyle,[80] an important trade unit on which was based the part of our Customary System that includes our present dry measures, as you will see in Chapter 4. In other words, this mural from over four thousand years ago illustrates, side by side, the two families of capacity standards that we still use in the United States today.

Neither of the standards pictured in the tomb of Hesy was originally Egyptian, and none of the measures shown there correspond directly to measures in the Egyptian system, which presumably were too well established to need memorialization. The main ancient Egyptian capacity units were the hon, known from a mathematical papyrus to have been $\frac{1}{300}$ of the cubic Egyptian cubit, and the hekat of 10 hon, $\frac{1}{30}$ of the cubic cubit;[81] these definitions imply a hon of 29.22 in³. Archeologically, the mean hon shown by surviving measuring vessels is 29.2 in³.[82] Rather, the tomb paintings were intended to show the main units the Egyptians had to reckon with in dealings with their major trading partners, or at least in those dealings that fell within the authority of this particular bureaucrat, Hesy, and to prove to the gods that he had done his job in maintaining these standards.

Similarly, there appears to have been a Sumero-Babylonian unit equal to $\frac{1}{150}$ of the Sumerian cubic cubit.[83] It's possible, therefore, that the gallon standard was not native to either place but was synthesized to facilitate trade between the two cultures.

The wine pint in Greece

Another ancient occurrence of a unit of the wine gallon standard comes from 20th century excavations at Olympia in Greece, which turned up a number of vessels specifically designed as capacity measures.[84] Their role as measures can be established with certainty from their distinctive forms and their inscribed rims, which prevented later alteration of the capacity. Most of the vessels exist only as fragments, but a few are whole enough to allow an accurate measure of their capacity, and these, together with the inscriptions, show the use of three separate standards for the measurement of three different categories of trade goods.

The oldest of the intact measures from Olympia, and the only intact representative of its category, was unearthed in 1954. Known as M6 in the catalogue, it dates from the late 5th century or early 4th century BC and is inscribed ΑΛ[Φ]Ι, an abbreviation for αλφιτον, which means "pearl barley" or "barley meal." Its capacity is 475 mL—that is, 28.99 in³. This is the wine pint again, corresponding to a gallon of 231.9 in³. Its use as a standard for barley aligns nicely with the 471 mL vase of Entemena noted earlier, since it's generally agreed that the Mesopotamian systems of measurement were based on barley.

As there was considerable classical Greek trade with Britain (known in the ancient world as a source of tin), it is quite possible that this Greek measure and its multiples provided the historical basis for our gallon. This would fit the probable inheritance of troy weight from Greek trade posited in Chapter 2.

The Hebrew gallon

It is generally agreed that the ancient Hebrews got their technology of measurement from Egypt and Mesopotamia; indeed, some of the names of their units (qab, seah, kor) come straight from the Babylonian (qa, sutu, kurru).[85] So it's not surprising that the gallon (but not the binary system we usually associate with it) also appears in the ancient Hebrew system of weights and measures. This gallon is the Hebrew hin, which appears in several places in the Old Testament.[86] The hin was $\frac{1}{6}$ of a measure that was called the ephah when used for dry goods[87] or the bath when used for liquids[88] and was half of a dry measure called the se'ah.[89]

There were actually two units named bath among the pre-exilic Hebrews, an ordinary bath of 6 hin and a royal bath (or bath of the sanctuary) exactly twice as large (i.e., equivalent to 12 hin).[90] This custom of defining a king's measure or king's weight to be exactly twice (occasionally four times) as large as the everyday unit of the same name was common in the ancient middle east.[91] Fragmentary remains of marked vessels have yielded reconstructed values for the royal bath of 45.3 to 46.6 liters, corresponding to a hin (gallon) of 230.4 to 237 in^3.[92] For the regular bath, two large amphorae give a value of 22.99 liters, which makes the hin 233.8 in^3.[93] From this it appears that the fourth of a hin specified in Exodus 29:40 and Numbers 15 is simply our quart and the qa found at Persepolis (page 57).

The size of the hin can also be estimated from the volume of the hen's egg, which is sometimes used in the Talmud as a standard of measurement. According to the Babylonian Talmud ('Er. 83a), the se'ah (2 hin) contained the volume of water displaced by 144 (a dozen dozen) hen's eggs of middle size, and the Jerusalem Talmud (Ter. 43c) effectively says the same thing by defining the egg as $\frac{1}{24}$ of a measure called the cab.[94] One Biblical archaeologist performed measurements on 100 hen's eggs of all sizes and found an average volume that puts the bath at 22.9 liters and the hin therefore at 232.9 in^3. The same researcher, taking another approach, performed an analysis of the molten sea of King Solomon's temple, whose dimensions are given in I Kings 7:26 and 2 Chronicles 4:5; he arrived at a calculated value of 22.794 liters for the bath, putting the hin at 231.83 in^3.[95]

Finally, there is the Talmudic tradition that the bath was the volume of a cube $\frac{1}{2}$ Hebrew cubit on a side, or in other words, that the bath was $\frac{1}{8}$ of the cubic Hebrew cubit.[96]

Historians[97] have put the value of one common variety of the Hebrew cubit at $\frac{6}{7}$ of the Egyptian royal cubit of 21.62–20.63 inches. The derivation of this odd-looking fraction ($\frac{6}{7}$) is actually quite straightforward. The Egyptian cubit was divided into 28 digits, i.e., 7 palms of 4 digits each, which made sense to the Egyptians for reasons that are interesting but have nothing to do with the present inquiry. The Hebrews took 6 palms of the 7, i.e., 24 of the same 28 digits, to make the more practical Hebrew cubit and referred to the longer Egyptian cubit as the "cubit and a handsbreadth" (Ezekiel 40:5 and 43:13).

Earlier I noted the very precise relationship between the cubic Egyptian cubit and the cubic Sumerian foot (see page 54). If we take the value of 20.624 inches for the Egyptian cubit that lines up exactly with the Sumerian foot of 330 mm, then the common Hebrew cubit ($\frac{6}{7}$ of the Egyptian cubit) is 17.68 inches and the corresponding bath ($\frac{1}{8}$ cubic Hebrew cubit) is 690.5 in^3. This is 3 hins of 230.2 in^3, which makes the bath defined in the Talmud exactly half the size we would have expected (6 hins). From this it appears either that the Talmudic tradition equating the bath with a cube half a Hebrew cubit on a side refers to a bath half the size of the ordinary bath, repeating again the derivation of the ordinary bath from the royal bath, or that somewhere along the line of transmission, the statement that the bath was half of a Hebrew cubit cubed became substituted for the statement that the bath was half a Hebrew cubic cubit. This latter interpretation would be exactly right if the bath were the royal bath of 12 hins.

It is interesting to compare the Hebrew bath calculated from the Talmudic definition (690.54 in^3) with the capacity of the Roman-era British Carvoran measure discussed on page 53 (average of two gaugings 691.08 in^3); the two volumes differ by less than a tenth of a percent. It's very unlikely that the Hebrew unit of measure actually traveled to England; but it's not hard at all to believe that the imperial accountants in Rome thought that they were working with two instances of the same standard.

The se'ah of 2 hin (gallons) seems to have been a common Semitic unit, because it survived into the 19th century as the gudda or cuddy of Arabia mentioned earlier, which measured 7.567 liters or 1.999 U.S. gallons.

Before leaving this topic, I must emphasize the universal view among scholars that Hebrew technology was almost completely derivative of systems used by the Egyptians and Babylonians, and therefore very unlikely to have been the historical source of our gallon. The relationship between our gallon and the ancient Hebrew hin is real, but it is a mathematical rather than historical one, evidence that they both came from the same bronze-age network of ancient systems of measurement that gave us the Northern foot and the troy pound.

Notes

[1] As this book is intended for readers in the U.S., I don't devote much space here to Imperial measure; for further information, see note on page 65 and the entry for Bushel in the Glossary beginning on page 139. For the benefit of my friends in Canada, however, I must dispel a misapprehension common there that the Imperial fluid ounce is the same as the U.S. fluid ounce. In fact, the Imperial fluid ounce ($\frac{1}{20}$ Imperial pint or $\frac{1}{160}$ Imperial gallon) is about 4 percent smaller than the U.S. fluid ounce ($\frac{1}{16}$ U.S. pint or $\frac{1}{128}$ U.S. gallon). For the exact figures, see the entry for Ounce in the Glossary.

[2] Many measuring spoon sets of recent manufacture sold in the U.S. equate the Customary tablespoon with 15 milliliters (mL). In fact, the Customary tablespoon is exactly half the U.S. fluid ounce, i.e., 0.90234375 in^3 or 14.78676478 mL. The difference (about 1.4 percent) is, of course, immaterial for cooking purposes, but the two measures are not identical.

[3] Scott, *Laws of the State of Tennessee*, 1:775.

[4] OED, s.v. "Gill."

[5] Lincoln, Johnson, and Northrup, *Colonial Laws of New York*, 1:99.

[6] Leaming and Spicer, *The Grants, Concessions, and Original Constitutions of the Province of New Jersey*, 508–09. As in many of the colonies, other measures were simply required to be "according to the standards of England" (ibid., 547).

[7] *Acts and Resolves of the Province of the Massachusetts Bay*, 1:49.

[8] George, Nead, and McCamant, *Charter to William Penn and Laws of the Province of Pennsylvania*, 239.

[9] *Acts of Assembly, Passed in the Province of New-York* (1719), 63; Lincoln, Johnson, and Northrup, *Colonial Laws of New York*, 1:555. Note the mention of "wine measure" prior to the act 6 Anne c. 27 (1706) discussed further on in this chapter.

[10] Bush, *Laws of the Royal Colony of New Jersey, 1703-1745*, 2:333.

[11] *Earliest Printed Laws of North Carolina*, 1:23–24. The barrel for pitch, turpentine, or tar was changed in 1784 to 32 gallons.

[12] *Statutes at Large of Pennsylvania*, 4:73.

[13] Alaska, Arkansas, Delaware, Idaho, Illinois, Kentucky, Massachusetts, Michigan, Minnesota, Mississippi, Missouri, Montana, North Carolina, Oregon, and Wisconsin.

[14] Many states specify, or did specify, barrels for various substances by weight rather than volume, the most common being the barrel of flour, defined throughout the U.S. as containing 196 pounds avoirdupois. Typically the older state laws gave the actual dimensions of these wooden barrels in addition to specifying the weights of their contents.

[15] "Whereas dayly experience sheweth that much fraud and deceit is practised in this colony by false weights and measures, for prevention whereof, Be it enacted that noe inhabitant, or trader hither, shall buy or sell, or otherwise make use of in trading, any other weights or measures than are used and made according to the statu[t]e of 12 Henry, VII, cap. V. (a) in that case provided:..." Hening, *Statutes at Large... of the State of Virginia* (1823), 2:89.

[16] More or less. In 1964 this definition was abandoned, and the liter is now officially just another name for the cubic decimeter.

[17] See Appendix A.

[18] Some historians have held that the weight equivalences in the statute of 1496 refer not to the wine gallon but to one of the other gallon standards current in England at the time, in particular, the Winchester gallon (Chapter 4); but this interpretation is physically impossible. See Appendix A for further discussion.

[19] *Excise to Treasury Correspondence* 3. Quoted by Berriman, *Historical Metrology*, 162.

[20] For a fuller account, see ADAMS, 42–43.

[21] Chisolm, *On the Science of Weighing and Measuring*, 68.

[22] For consistency, I've used citations from the last and presumably best of the compilations of the old statutes, the one authorized by George III. Sometimes the earlier printed versions numbered statutes a bit differently; for example, this statute is referenced in older compilations as 2 Henry 6 c. 11, so that's the way you will see it cited in earlier histories. I've provided a set of tables to help sort this out in Appendix B.

[23] In the original:

Item combien qen auncien temps fuist ordeinez & loialment usez qe toneux pipes tercians hoggeshedes de vyn de Gascoigne, barelles de Harank & danguilles & buttes de Samon, veignant par voie de merchandise en cest terre hors des estraunges pais & auxi faitz en mesme la terre, serroient de certein mesure cest assavoir, le tonell de Vyn de xij^{xx} & xij galons, le pipe de vj^{xx} & vi galons, la tercian de iiij^{xx} & iiij galons, le Hoggeshede de lxiij galons, le Barell de Harank & danguilles de xxx galons pleinement pakkez, le butte de Samon de iiij^{xx} & iiij galons pleinement pakkez; nientlemains par ymaginacion & subtilite ount jatard estee faitz tieux vesseux de pluis petite mesure, a graunde perde & desceit au Roy & de son poeple, dount en ceste parlement fuist priez especialment de remedie:...

See Appendix B for notes on the treatment of these translations.

[24] *tonell de Vyn sil ne conteigne del mesure dEngleterre xij^{xx} & xij galons, le Pipe vj^{xx} vj galons, & ensy solonc lafferaunt le tercian & le Hoggeshede de Vyn de Gascoigne...*

[25] The practice of recording English laws in Norman French ended around 1487, and by the time this statute of 1483 appeared, the laws were drawn up in parallel English and French versions that sometimes don't track each other exactly. It's hard to tell which version is the original, so rather than attempt my own synthesis here, I have deferred to the English version given in Pickering (ca. 1763).

[26] *par butte Barell dymy Barell ou ascun autre vesseau devant quil soit vieu, sinon mesme le Butte teigne and conteigne quatre vins & quatre galons, le Barell quarant deux galons, Et le demy barell xxj galons, bien & foialment pakkez...*

[27] *par Barell di. Barell ou firkyn, sinon le Barell conteigne xlij galons, le demy barell & Firkyn sononq mesme la rate...*

[28] TATE'S, 62.

[29] The standard in Paris appears to have been a little larger; ADAMS (84) cites a value for the pre-metric Paris pinte of 58.08 in^3, which is one fourth of 232.32 in^3.

[30] TATE'S, 484.

[31] TATE'S, 151.

[32] TATE'S, 125.

[33] In continental Europe, Roman and Phoenecian standards are equally important and account for many other premetric survivals.

[34] The description of the Carvoran Measure was originally reported in 1916 by F. Haverfield ("Modius Claytonensis: The Ro-

man Bronze Measure from Carvoran"). See also A. E. Berriman, "The Carvoran 'Modius.'" The 1948 measurements and further discussion can be found in SKINNER, 69–72. Haverfield also cites a discussion by "Mr. W. Airy of Greeenwich in the monthly *Review of the Incorporated Society of Inspectors of Weights, etc.,* Dec. 1915, p. 243."

[35] Such leveling bars go back at least to Greek models of the sixth century B.C. (SKINNER, 61). They are found on two well-preserved examples from Herculaneum and Pompeii, and the design continued to be used on measures in Northern Italy into the 19th century of our era (Haverfield, "Modius Claytonensis," 90–92).

[36] Unfortunately, this definition shifted about a bit over time. The Weights and Measures Act of 1824 (5 George 4 c. 74) that created Imperial measure defined the Imperial gallon as the measure that contains "10 imperial pounds weight of distilled water weighed in air against brass weights with the water and the air at a temperature of 62 degrees of Fahrenheit's thermometer and with the barometer at 30 inches" and set the bushel of 8 gallons at 2218.192 in^3, which puts water at 252.45785757... grains/in^3; but in 1889, an Order in Council changed the volumetric definition of the Imperial bushel to 2219.704 in^3, corresponding to a weight for water of 252.28589037... grains/in^3. More recent laws put the Imperial bushel at the equivalent of 2219.35546... in^3 and abandon the definition based on a weight of water, just as the SI has abandoned the definition of the liter based on a weight of water. The slight differences among these definitions are negligible in the context of this discussion, and I note them only to make clear the basis for various calculations made here and elsewhere in the book. For consistency, I've used the density for water set in 1889 (still current when the Carvoran measure was gauged in 1948) as the basis for all such calculations unless otherwise noted. For further information about Imperial measures, see the entries for Bushel, Gallon, and Ounce in the Glossary beginning on page 139.

[37] The extensive literature on this subject will be found under the headings "Farnese congius" and "congius of Vespasian."

[38] Ancient authorities differ on whether the cubic Roman foot held 80 pounds of water or 80 pounds of wine. The difference is small enough to have fallen below the threshold of significance in many cases.

[39] Petrie, *Ancient Weights and Measures,* pl. 53. The set was probably for weighing gold coins called scrupula or scrupular aurei; Hussey (*Essay on the Ancient Weights and Money,* 124), reports "scrupular aurei" with values of 17.52 troy grains (computed by M. Letronne based on an average of 27 coins) and 17.24 grains (based on four coins weighing 1, 2, 3, and 4 scrupula in the British Museum). See also Humphries, *Coin Collector's Manual,* 1:273–74, and Smith, *Dictionary of Greek and Roman Antiquities,* s.v. "Aurum." This early coin is not to be confused with the later aureus.

[40] Our Customary foot was also traditionally divided into 16 fingers or finger-breadths (look up *finger* in any dictionary of English). The Greeks used digits only, dividing their foot into sixteenths. The Romans added an alternative division into inches (called "thumbs" in most of the Germanic traditions) and made foot rulers that were divided into 12 inches on one edge and 16 digits on the other.

[41] Peet, "Mathematics in Ancient Egypt," 415.

[42] Parker, *Demotic Mathematical Papyri,* 10.

[43] Griffith, "The Metrology of the Medical Papyrus Ebers," 393; Sloley, "Ancient Egyptian Mathematics," 111–17; Peet, "Mathematics in Ancient Egypt," 413.

[44] The average value of the cubit found in the six most important 4th–6th dynasty pyramids and temples (Petrie, *Pyramids and Temples of Gizeh,* 179).

[45] Isaac Newton reviewed much earlier estimates and put the Egyptian cubit at 1.719 feet, which is 20.628 inches (Glazebrook, "Standards of Measurement," 19). Drovetti (*Lettre a M. Abe Remusat,* n.p.) reported four exceptionally well preserved Egyptian cubits in Memphis averaging 52.351 ±0.039 cm (20.6105 ±0.0155 in).

[46] Thureau-Dangin, "Numération et Métrologie Sumériennes," 133). I believe that SKINNER (41) is wrong in considering the Sumerian foot a variant of the Northern foot.

[47] As far as I can tell, neither Petrie nor Thureau-Dangin was aware of this relationship, a lapse that in the absence of calculators with cube root keys is certainly understandable and demonstrates that these two authorities were not influenced by each other's work in this regard.

[48] The Austrian fuss was officially $\frac{1}{6}$ of a klafter of 1.896484 meters.

[49] I'm sorry to say that the quarter-century delay in completing this history means that I'm not the first person to have published this observation. In finishing up the book, I came across a work by Daniel McLean McDonald, a British industrialist who apparently created an institute in order to publish a collection of his papers titled *The Origins of Metrology* (Cambridge: McDonald Institute for Archaeological Research, 1992), on page 8 of which he notes that 10 gallons (gallon = 230.4 in^3) is the cube of a foot of 13.20770901 inches (McDonald gives all of his figures to 10 places). McDonald has apparently never heard of the Northern foot and is incorrect, in my opinion, in considering this foot to be the Sumerian foot; in fact, it seems to me that he is incorrect in most of his conclusions. But while I had already noticed the relationship between the gallon and this cubic foot back in the mid-1980s, McDonald gets the credit for publishing first.

[50] TATE'S, 256.

[51] [Chisolm], *Tenth Annual Report,* vii.

[52] $\tilde{q}$ null march^{a}unt nautre persone mette ascun harank au vendre per barell dī barell ou firkyn, sinon mesme Barell conteigne xxxij galons, le dī Barell & Firkyn solonq mesme la rate;...

[53] Holinshed, *Chronicles,* 122 (recto).

[54] CONNOR, 174.

[55] That is, $26\frac{2}{3}$ Imperial gallons of 277.274 in^3 (as defined at the time) is 32 wine gallons of 231.0617 in^3.

[56] Grimke, *Public Laws of the State of South-Carolina,* 209.

[57] Prince, *Digest of the Laws of the State of Georgia,* 816.

[58] Potter, *Laws of the State of North-Carolina,* 1:478–79.

[59] Toulmin, *Digest of the Laws of the State of Alabama,* 360.

[60] Thureau-Dangin, "L'u, le Qa et la Mine," 91, n. 2.

[61] Schmidt, *Persepolis II,* 108–09.

[62] Cunliffe, *Europe between the Oceans,* 227.

[63] "Report of Diggings in Silbury Hill," 216–17; *Encyclopædia Britannica* (1942), s. v. "Measures and Weights, Ancient," 143.

[64] *Inductive Metrology,* 108.

[65] *Megalithic Sites in Britain,* 43. See also Thom and Thom, *Megalithic Remains in Britain and Brittany,* 39–43, and Thom, "The Geometry of Cup-and-Ring Marks," 77–87.

[66] Petrie, *Ancient Weights and Measures,* 46–47.

NOTES

[67] ADAMS, 99 and 186–87. Corn was a generic term back then for all grains, not just maize, which was known as Indian corn.

[68] JOHNSON'S, 492; MARTINI, 277.

[69] UN (1955), 35, 88.

[70] SKINNER, 41.

[71] JOHNSON'S, 484.

[72] TATE'S, 352.

[73] JOHNSON'S, 483. The reported size of 1.999 U.S. gallons appears to have been calculated from the metric figure (7.567 liters is 1.99898+ U.S. gallons or 2 gallons of 230.88 in^3). There is no historical reason to believe that the unit came from England (the Imperial gallon would have been used), and it is very unlikely that it was borrowed from the U.S.; if it had been, *Johnson's* would have given it as 2 gallons. The metric figure cited argues against this, too; a metric conversion from 2 gallons of 231 in^3 would have given 7.571 liters, not 7.567.

[74] SKINNER, 5–6.

[75] From Quibell, *Tomb of Hesy*, pl. 17.

[76] Petrie, *Ancient Egypt* (1915), Part I, 39.

[77] Quibell, *Tomb of Hesy*, pl. 13.

[78] Quibell, *Tomb of Hesy*, 26.

[79] Petrie, *Ancient Weights and Measures*, 35; Petrie, *Measures and Weights*, 9.

[80] Petrie, "Measures and Weights, Ancient," 173. See also SKINNER, 45–46. The kotyle was used for trade in wine and olive oil. It should be noted that Petrie and Skinner are not responsible for the identification made here between these standards and units of the Customary System.

[81] Griffith, "Notes on Egyptian Weights and Measures," 406; Gillings, *Mathematics in the Time of the Pharaohs*, 163.

[82] Petrie, "Weights and Measures, Ancient Historical," 484. The hon survived in Egypt into the 19th century A.D. as the roub-kaddah of 477.8 cc or 29.1571 in^3 (JOHNSON'S, 483).

[83] That is, if the cubit is reckoned as 495 mm, about 808.6 mL. The evidence includes, first, an inscribed but fragmentary alabaster vessel from the palace of "Evil-Mérodach, roi de Babylone, fils de Nabuchodonosor" with a capacity corresponding to a unit of about 0.81 liters (Thureau-Dangin, "La Mesure du *qa*," 24–25, also illustrated in Scheil and Legrain, *Textes Élamites-Sémitiques*, 60), and second, an Old Babylonian jar implying a unit of 806.87 mL (Postgate, "An Inscribed Jar from Tell al Rimah," 73); the average of these two figures is 808.435.

[84] Schilbach, "Massbecher aus Olympia," 323–56. I am grateful to Professor Ronald S. Stroud of UC Berkeley for calling this study to my attention.

[85] Lewy, "Assyro-Babylonian and Isrealite Measures of Capacity," 69.

[86] Exodus 29:40; Ezekiel 45:24, 46:11, and 46:14.

[87] Exodus 16:36; Ezekiel 45:11, 45:13, and 46:14.

[88] Ezekiel 45:11 and 45:14; 2 Chronicles 2:10.

[89] Genesis 18:6; 1 Samuel 25:18; 1 Kings 18:32; 2 Kings 7:1 and 7:18.

[90] For example, Hussey, *Essay on the Ancient Weights and Money*, 183ff.

[91] For example, *Jewish Encyclopedia*, 484.

[92] Scott, "Weights and Measures of the Bible," 29.

[93] Ibid., 30.

[94] *Jewish Encyclopedia*, s. v. "Weights and Measures," 487.

[95] Zuidhof, A., "King Solomon's Molten Sea and (π)," 183. The reader interested in puzzles may spend many a happy hour trying to solve the riddle of the size of the Molten Sea; in addition to Zuidhof, see Wylie, "On King Solomon's Molten Sea," and Scott, "Weights and Measures of the Bible," 26.

[96] Petrie, "Weights and Measures (Ancient Historical)," 485.

[97] SKINNER, 39; and others.

Chapter 4

The Winchester bushel

And thus much briefly of liquid measures, wherein yf I have beene more long & tedious then thou peradventure diddest looke for at the first, yet the benefite gotten thereby shall, I hope, countervaile the travaile in reading of the same. And as I have dispatched my handes in this sort of the liquide, now it resisteth that I doe the lyke with the drie measures, & then shall that little Treatise have an ende, whereof I spake before....

Holinshed's *Chronicles* (1577)

Figure 4.1: Exchequer standard Winchester bushel of Henry VII

As you saw in the last chapter, our U.S. liquid gallon, a.k.a. the wine gallon, is a standard of capacity whose history can be traced back more than four thousand years. Though based upon a weight of wheat, the wine gallon came to be used for almost everything but grain, continuing here in the U.S. as a measure strictly for liquids.

Many centuries ago, before the Norman invasion in 1066, the English adopted a different standard for grains and fruits, and that's what we use in the U.S. today for the same purpose. This old standard is known historically (and still known in U.S. Department of Agriculture regulations) as the Winchester bushel.

The term *Winchester measure* goes back to King Edgar the Peacable (AD 959–75), who ruled that "the measure of Winchester should be the standard for his realm" (*et una mensura, sicut apud Wincestrum*).[1] But there's no doubt that its history as part of a specific set of standards goes back farther than that.

Along with its smaller subdivisions, the U.S. dry peck, dry gallon, dry quart, and dry pint, the Winchester bushel is still our Customary capacity standard for the measurement of dry substances. Nowadays, due to the automation of weighing devices, we commonly see dry goods sold commercially by weight rather than capacity, but the old Winchester system is still the standard by which we sell fruits and vegetables (using the quart berry basket, for example).

It's important occasionally to remember that in the set of traditions from which Winchester measure comes, grains and salt are measured striked (or struck), that is, leveled off using a rod-like object called a *strike*, but fruits and vegetables are sold heaped up. The role that heaping has played in the development of capacity standards is a subject unto itself; the important thing to note here is that when measures are leveled off, the shape of the measuring vessel doesn't make any practical difference, but when goods are sold heaped, it very much does: the wider the measure, the more the heap contains on top of what the measure would have held striked. This is why many state laws specify the dimensions of the containers for fruits and vegetables and not just their capacities.

The history of the Winchester measures demonstrates some basic principles underlying all ancient systems of measurement—most importantly that measures of capacity are simply related to the weights of the substances being measured and that

Chapter 4. The Winchester bushel

systems of weights and measures are arranged in a way that simplifies the calculations needed to plan the shipping and storage of basic trade goods. These considerations will also throw some light on the origins of our Customary shorter measures of length (inch, foot, and yard), as discussed further in Chapter 6.

The Winchester system

The units of Winchester measure are:

	Unit	Gallons	Bushels
	pint	$\frac{1}{8}$	$\frac{1}{64}$
×2 =	quart	$\frac{1}{4}$	$\frac{1}{32}$
×4 =	gallon	1	$\frac{1}{8}$
×2 =	peck	2	$\frac{1}{4}$
×2 =	demibushel	4	$\frac{1}{2}$
×2 =	bushel	8	1
×8 =	quarter	64	8
×4 =	chaldron	256	32

As with wine measure, units of Winchester measure form essentially a binary series, but the pint, quart, and gallon of the Winchester measure are substantially larger than the corresponding units of wine measure—another demonstration of the difference between a system and a standard. By law in the U.S., the Winchester bushel contains 2150.42 in^3, exactly.[2] This gives a Winchester gallon of 268.8025 in^3, which is about one-sixth larger than the U.S. liquid gallon (the old wine gallon) of 231 in^3.

The Winchester bushel has always been the legal standard of dry measure in nearly all of the United States, first appearing by statute in a Plymouth Colony law dated 1 July 1633, just 13 years after the Pilgrims landed.[3]

> It was decreed that ye new bushell, being a seald bushell brought out of England of Winchester measure should be alowed and no other, and all other measures to be brought into the constable to be made conformable to the same, and so to be sealed by him with the seale appoynted for that end and this to be done by the last of this present month. But notwithstanding that all former bargains and sales that were made before this day, they are to be fulfilled by old measure.[4]

In a more organized set of Plymouth Colony laws passed in 1636, the section headed "Weights and Measures" consists entirely of the following two sentences:

> That one comon standard to be used by all for weight and measure. And that according to Winchester which is the standard of Engl.[5]

In what later became the states of New York, New Jersey, and Pennsylvania, a 1665 decree of the Duke of York (who afterward became King James I and had almost absolute power in those colonies) required "That the high Constable in each Riding shall provide at the Publique charge Severall Standards of weight and Measures," and

> That there be one Bushel one Peck, and one halfe Pecke to be fitted to winchester measure in England; and Measures for Liquids as the Ale quart Wine quart, wine Pint, and halfe Pinte, And that there be one Ell, and one yard, that all and each may be according to the General Custome of England...[6]

Winchester barrels

Several colonial laws defined barrels for meat and grain on the Winchester standard. A 1631 Virginia law ordered

> That a barrell of corne shall be accounted five bushells of Winchester measure, that is to say, 40 gallons to the barrell, and that the comissioners, for the mounthlie corts throughout the colony, doe take order and see that sealed barrells are made and sealed with this seale as in the margent, [VG] which seale they are to keepe, and upon request to seale such barrells, and bushells, as shall be brought unto them.[7]

Typically for colonial law, the pushment for anyone using an unsealed barrel or bushel was a fine and some time in the pillory. A slightly revised version of the same law specifying penalties for second and third offenders was issued in 1632,[8] and the same definition of the barrel as 5 Winchester bushels appears again in laws of 1641,[9] 1642,[10] 1657,[11] and 1661.[12] An 1814 Virginia law defining the barrel for salt as 5 bushels[13] suggests the existence of the unit in England, even though it appears nowhere in the English statutes.

A 1705 Virginia law added a barrel for pork, beef, tar, and pitch at $31\frac{1}{2}$ gallons "of Winchester measure."[14] Both barrels appear to have stayed on the books in Virginia until 1849.[15] The barrel of $31\frac{1}{2}$ Winchester gallons, probably created by confusion with the barrel of $31\frac{1}{2}$ wine gallons that appears in a number of English statutes (Chapter 3), was also specified for barrels of pork and beef by New York laws of 1692 and 1693.[16]

Louisiana, possibly due to its origin as a French colony, defined the barrel in a law of 1814 as $3\frac{1}{4}$ bushels and then redefined it in 1855 as $3\frac{1}{2}$ bushels.[17] Current Louisiana law defines the barrel as 6451.26 in^3 "which approximately represents the contents of three bushels." In fact, 6451.26 in^3 is exactly 3 bushels of 2150.42 in^3. The same statute defines a unit called a sack that is half of this barrel (i.e., $1\frac{1}{2}$ bushels).

Michigan, North Carolina, and West Virginia statutes recognize a dry barrel for fruits, berries, and vegetables of 105 Winchester quarts (7056 in^3). This same unit is listed in the U.S. Bureau of Standards publication *Units and Systems of Weights and Measures*[18] as "barrel, standard for fruits, vegetables, and other dry commodities, except cranberries, 7056 cubic inches, 105 dry quarts, 3.281 bushels, struck measure."[19] Cranberries get their own barrel of "5826 cubic inches, $86\frac{45}{64}$ dry quarts, 2.709 bushels, struck measure." These strange definitions resulted from older state laws giving the actual dimensions of barrels as constructed, a common alternative to giving their capacities in units of measure.[20]

Winchester measure for liquids

While Winchester measure in the U.S. eventually came to be used exclusively for dry substances, his-

Figure 4.2: Exchequer standard Winchester gallon of Henry VII

torically it was also used for liquids, in particular beer and ale.

For example, a 1631 domestic paper of King Charles I is summarized as follows:

> Account of the Constables of the Duchy liberty Strand Westminster, of money received by them under warrant of Sir William Slingsby, being forfeitures to the King for selling less than a full Winchester quart of beer or ale for a penny.[21]

In 1701, the Officers of the Excise complained to the Lords of the Treasury that "malt drink of excessive strength" was hurting taxable consumption because

> the manufacturers, artificers, farmers, labourers, and travellers, who were the great consumers of exciseable beer and ale, generally measured their drinking by the money they could afford, and if they drank common beer or ale would drink a full Winchester quart for 2*d.* or $2\frac{1}{2}d.$, whereas they drank no more than one pint of this extraordinarily strong drink, for the same or more money.[22]

According to a recent history of Halesworth, England,

Chapter 4. The Winchester bushel

It was the duty of the "ale founders and tasters" to test the beer sold at the various establishments and to see that the correct measure was given. In 1651 Nathaniel Chilston and Walter Winston were chosen as ale founders and tasters in Halesworth. They were provided with "One Winchester quart and one pint of pewter, a set of weights ranging from half quarter of an ounce to eight pounds, a book of directions for officers, one brand to mark the measure and a pair of brass scales."[23]

As one would expect, use of Winchester measure for liquids was in some places carried over into U.S. colonial practice. A law of the Colony of New Plymouth enacted some time between 1623 and 1636 mandated

> That none be suffered to retale wine strongwater or beere either within doores or without except in Inns or Victualling howses allowed. And that no beere be sold in any such place to exceed in price two pence the Winchester quart.[24]

Parallel use of the Winchester standards for both liquid and dry measures is nicely demonstrated in the foundational set of statutes known in Pennsylvania history as The Great Law, passed by the Assembly under William Penn in 1682, which contains the following two adjacent paragraphs:

> **Chapter 38** And to prevent Exaction in Publick houses be it further Enacted by the Authority aforesaid that all Strong beere and ale made of Barley Malt shall be Sould for not above two pennys Sterling a full Winchester Quart and all Beere made of Mollassus Shall not Exceed one penny by the Quart.
>
> **Chapter 39** And to prevent fraude in Measures and to reduce all forreigne Measures here to the English Standard be it further Enacted by the Authority aforesaid that the Measures of this Province shall be according to the Standard of Weights and Measures in England that is to say a Bushell Shall Containe Eight Gallons according to the Winchester Measure and all Weights to be averdupois which hath Sixteen Ounces to the pound within three Months after the first Session of this Assembly.[25]

The placement of these two statutes together in the list leaves no doubt that the same Winchester standards were used in Pennsylvania at that time for both liquids and dry substances.

By the end of the colonial period, however, use of Winchester measure for liquids seems to have entirely disappeared, that role being taken over exclusively by wine measure (Chapter 3).

Chemists and readers of fiction set in the early 19th century will occasionally encounter references to a liquid measure called the Winchester quart that has nothing to do with the Winchester quart still used as a dry measure in the U.S. With the abolition of all but Imperial measure in England in the early 19th century, the term "Winchester quart" floated free from its previous meaning and came to be applied to a kind of bottle used in chemistry labs. This bottle originally held 80 Imperial ounces (2 Imperial quarts) or 2.273 liters, finally settling down to a capacity of $2\frac{1}{2}$ liters, where it can still be found in some catalogues of laboratory ware. Apparently the name "Winchester pint" was used for a while in a similar way.[26] The names "Winchester quart" and "Winchester pint" for the laboratory bottles have never been common in the U.S., because the real Winchester quart and pint continued in use here for the sale of fruits and vegetables, as they do today.

Modern definition of the Winchester bushel

As in the case of the wine gallon, there were two phases of definition for Winchester measure: an 18th century scientific definition in terms of cubic inches, and statutes hundreds of years earlier that defined the same units in terms of weights of wheat. We'll start with the later definition because that's the one that gave us the legal value we still use in the U.S.

The first definition of the Winchester bushel in terms of cubic inches[27] appeared in 1696–97 toward the end of a statute of William III (8 & 9 William 3 c. 22 §45) regarding the sale of malt.

> And to the End all his Majesties Subjects may know the Content of the Winchester Bushell whereunto this Act refers and that all Disputes

and Differences about Measure may be prevented for the future It is hereby declared That every round Bushel with a plain and even Bottom being eighteen Inches and a Halfe wide throughout & Eight Inches deep shall be esteemed a legal Winchester Bushel according to the Standard in His Majesties Exchequer.[28]

The cylinder thus defined contains 2150.42017 ... in^3, and our statute definition of 2150.42 in^3 is simply this value to seven places of precision (i.e., 2150.420). As indicated by the phrase "according to the Standard in his Majesty's Exchequer," however, the standard is much older than this definition. The Exchequer still contains standards of the Winchester bushel and Winchester gallon dating to Henry VII (1496), and the average of the two shows a bushel of 2146.125 in^3.[29] See Figure 4.6 (page 81) for a contemporary picture of their making.[30] Exchequer bushel and gallon standards of Elizabeth I show a Winchester bushel of 2154.92 in^3,[31] and the average of these two sets gives us a figure for the original Winchester bushel of 2150.52 in^3, virtually identical to our current legal definition.

Original definition of the Winchester bushel

It probably won't come as a surprise at this point to learn that our Customary standard for dry measure can be traced back almost as far as our standards for land measure, liquid measure, and troy weight. I must acknowledge, however, that establishing the early history of Winchester measure in England is not a simple task. I am confident of the reconstruction offered here, but at least one key question will not be completely resolved until the subject of grain weights is settled in Appendix A.

What appears to be the oldest surviving definition of Winchester measure is contained in language attached to some copies of the statute known as the *Assisa Panis et Cervisie* or Assize of Bread and Ale. It is of uncertain date, but tradition assigns it to 51 Henry 3 (1266). The relevant passage (translated from the Latin) says:

By consent of the whole realm of England, the measure of our Lord the King was made; that is to say, that an English penny, called a sterling, round and without clipping, shall weigh thirty-two grains of wheat in the middle of the ear; and 20 pennies make an ounce; and twelve ounces make a pound; and eight pounds make a gallon of wine; and eight gallons make a bushel of London; which is the eighth part of a quarter.[32]

Similar language appears in a somewhat later statute, the *Tractatus de Ponderibus et Mensuris* (Treatise on Weights and Measures), also of unknown date but traditionally referenced as 31 Edward 1 (1301).[33] There are several extant versions of the Latin text. The one cited in 18th century editions of the *Statutes of the Realm* translates as:

By the ordinance of the whole realm of England, the measure of our Lord the King was made, that is to say, that the penny called sterling, round and without clipping, shall weigh thirty-two grains of wheat in the middle of the ear. And the ounce shall weigh twenty pence. And twelve ounces make a pound of London. And twelve and a half pounds make a stone of London.[34] And eight pounds of wheat make a gallon. The pound contains twenty shillings. And eight gallons make a bushel of London.[35]

The manuscript chosen for the official edition of the *Statutes of the Realm* produced by order of George III is titled *Assisa de Ponderibz et Mensuris*[36] and translates as:

By the ordinance of the whole realm of England, the measure of our Lord the King was made, that is to say, that the English penny called sterling, round and without clipping, shall weigh thirty-two grains of wheat in the middle of the ear. And the ounce shall weigh twenty pence. And twelve ounces make a pound of London. And eight pounds make a gallon of wine. And eight gallons of wine make a bushel of London. And eight bushels make a quarter of London. And twelve and a half pounds make a stone of London.[37]

If you compare these texts with the 1496 statute defining the wine gallon and the troy pound on page 45 (Chapter 3), you will find them very

Chapter 4. The Winchester bushel

similar. In particular, both the statute of 1496 and the *Tractatus* of two centuries earlier set up basically the same series of weight units, and both define a gallon as eight pounds of wheat. They both refer to the same traditional system of *relationships*—but they do not refer to the same *standard*. The wine gallon defined in the statute of 1496 as 8 troy gallons of wheat cannot be the Winchester gallon; the units are of distinctly different sizes, and 8 troy pounds per Winchester gallon is not a physically possible weight for wheat of good or even average quality (this is explained in detail in Appendix A). It follows that the Winchester measures must be based on upon a different standard of weight.

The tower pound

Alongside the ancient 480-grain troy ounce, whose history we recounted in Chapter 2, there also existed a 450-grain ounce. The larger unit was traditionally associated with gold, and the smaller one with silver, the parallel use of the two ounces being well attested in medieval Arabic weights.[38] England's currency was based on silver, so the coinage standard became the pound of 12 450-grain ounces. This unit was known in England as the tower pound, because the standard weights were kept in the Tower of London; and because the coins were called sterlings, this monetary unit was also called the pound sterling, a name it has kept to this day, though as a unit of account it stopped representing the value of a tower pound of silver centuries ago.[39] Official references to the tower pound by name go back at least to 1444, when it was prayed in Parliament to coin more "Half Penyes and Ferthings" from "every pound weight of the Tour."[40] The statute 9 Henry 5 stat. 2 c. 6 (1421) was undoubtedly referring to tower weights when it proclaimed

> That all the money of gold and silver that shall be made at the Tower of London and at Calais, or elsewhere within the Realm of England, by authority royal, shall be made of as good allay, and just weight, as it is now made at the Tower.[41]

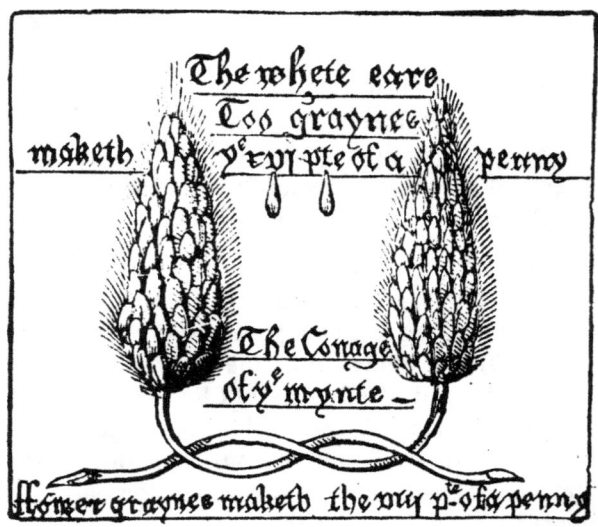

Figure 4.3: "The whete eare. Too graynes maketh y^e xvi pte of a penny. The conage of y^e mynte—ffower graynes maketh the viii p^{te} of a penny." 1841 woodcut copy of an illustration (now lost) from the time of Henry VII

The dual system of troy and coinage (tower) weights was finally abolished by an Exchequer verdict relating to coinage dated 30 October 1527 (18 Henry 8) that contains the following passage:

> Where as hertofore the Marchaunte paid for Coynage of every Pounde *Towre* of fyne gold, weying xi *oz.* qurt. ii *s.* vi *d.* [that is, for turning every tower pound of gold, weighing $11\frac{1}{4}$ troy ounces (5400 grains), into coins, the merchant paid 2 shillings and 6 pence]. Nowe yt is determyned by the Kingis Highnes and his said Councelle (that the forsaid pounde *Towre* shall be no more used and occupied) but Alman[ner] of Goldes and Sylver shal be wayed by the Pounde Troye, which maketh xii *oz.* Troye, which exceedith the Pound *Towre* in Weight iii quarters of the *oz.*[42]

This explicitly puts the tower pound at exactly $\frac{15}{16}$ the weight of the troy pound, that is, 5400 grains instead of 5760. The English silver penny did in fact weigh $22\frac{1}{2}$ grains rather than the 24 grains of the troy pennyweight, a $\frac{15}{16}$ ratio.

The tower pound must have been long abandoned for all uses other than coinage by the time this policy was adopted, as there appears to be no other official reference to it, but its former existence was confirmed by the finding in 1842 of a pound weight standard of 5391 grains in the Pyx Chamber in the cloister of Westminster Abbey.[43]

According to a chapter that "treateth of a waight called Tower Waight" in a 1606 collection of tables by a London goldsmith,

> This is a waight which hath been used in England from the beginning in the Kinges Mintes till of late yeeres and deriued from the Troy waight of 12 oz., by which the merchant bought his gold and siluer abroad, and by the same deliuered it into the Kinges Mintes, receyvinge in Counterpoise, by Tower waight for Troy waight, which was the Prince's prerogative, [who] gayned thereby 3 quarters of an ounce in the exchange of each pound waight, beinge converted into moneyes; beside the gaine of coyninge, which riseth to greate reuenewe, makinge of euery xxx^{li} waighte Troies (being a journey of quoyned moneyes) 32^{li} wayhtes Toweres.[44]

In other words, merchants bringing in gold and silver from abroad and needing it coined into English money traded every 480-grain troy ounce of metal for a 450-grain ounce of coins, or every 24-grain pennyweight of silver for an actual $22\frac{1}{2}$-grain sterling penny.

But this explanation doesn't seem to help much in relating the definition in the *Tractatus* to the surviving standards. If 8 troy pounds of wheat occupy a wine gallon of 231 in^3 (which they quite accurately do; see Appendix A), then 8 tower pounds of that wheat occupy 216.5625 in^3—nowhere close to the Winchester standard gallon of 268.8025 in^3 or any other surviving English capacity measure.

The key to this puzzle comes further on in the *Tractatus de Ponderibus et Mensuris*. After a long list of weights associated with wool, hides, and other goods, it continues:

> And it is to be known that every pound of pennies [i.e., coins] and spices as electuaries [medicines] consists of a weight of only twenty shillings. But the pound of all other things consists of twenty-five shillings. Truly the ounce of electuaries consists of 20 pennies, and the pound contains twelve ounces. But in other things, the pound contains fifteen ounces, the ounce of this [likewise] being twenty pennies in weight.[45]

The pound of 15 tower ounces, that is, $12 \times 450 = 6750$ troy grains, is called the libra mercatoria or tower merchant's pound.[46] The earliest statute reference to the merchant's pound appears to be a law of the Saxon king Æthelred Unrædy (ruled 969–1016) that commanded "those who have the keeping of the ports, and the collecting of the customs on goods, that, under the pain of my displeasure, they collect my money by the pound of the market; and that each of these pounds be so regulated and stamped as to contain 15 ounces."[47]

You already saw this pattern of light and heavy pounds in our discussion of troy weight:

12 ounces make the pound for small or precious things

15 ounces make the pound for everything else

In the case of troy weight, the pound of 15 troy ounces corresponded to the Hanseatic unit known as the Pound of Cologne (page 29ff.). Note the $\frac{4}{5}$ ratio between 12 and 15 that's characteristic of the pattern; this will become important later.

From this it's clear that the reason we originally calculated a volume that's way too small from the definition of a gallon as weighing 8 pounds of wheat in the *Tractatus* is because we were using the wrong pound—the pound of 12 tower ounces used for coinage rather than the pound of 15 tower ounces used for most other things. Since wheat is one of the things weighed by the 15-ounce tower merchant's pound of 6750 grains, it follows that 8 pounds of wheat will occupy about 270.7 in^3, which is just about right for the Winchester gallon. Let's put it this way: If a current U.S. dry gallon of wheat weighs 8×6750 troy grains, then that wheat weighs 61.71 avoirdupois pounds per U.S. bushel. As noted earlier, actual U.S. No. 1 bread wheat weighs about 61.0 lb/bu. Hard winter wheat consistently weighs slightly more than that.

Chapter 4. The Winchester bushel

That this interpretation of the somewhat convoluted formula preserved in the official copies of the *Tractatus* is the correct one is confirmed by another copy of the same statute from the time of Edward I that translates as follows:

> By ordinance of the whole realm of England, the measure of our lord the King was made, namely the English penny called sterling, round and unclipped, shall weigh thirty-two grains of wheat in the middle of the ear. And an ounce ought to weigh twenty pennies. And fifteen ounces make a pound of London. And eight pounds of wheat make a gallon of wine. And eight gallons of wheat make a bushel of London, that is, the eighth part of a quarter. Twelve pounds and a half make a stone of London. A sack of wool ought to weigh 28 stones [more on this in Chapter 5]...
>
> It is to be known that a pound of pennies, spices, confections as electuaries is equal in weight to twenty shillings. But a pound of other things weighs 25 shillings. In electuary confections a pound contains twelve ounces, and an ounce is equal in weight to 20 pennies [in both cases].[48]

A virtually identical version of the same statute has been kept for centuries by the Borough of Northampton.[49] The construction "fifteen ounces make a pound of London, and eight pounds of wheat make a gallon of wine" confirms that the gallon is the Winchester gallon, because no other English gallon standard contains that actual weight of wheat (see Appendix A).

I would be remiss, however, if I did not mention a less likely but truly elegant alternative interpretation argued at length by John Quincy Adams in his *Report to Congress on Weights and Measures*. According to Adams, the language in the *Tractatus* means not that the gallon contains a quantity of wheat weighing eight tower merchant's pounds but rather that the gallon when filled with wine *weighs the same as* eight tower merchant's pounds of wheat. This would make the gallon occupy a little more than 216 in^3, as noted earlier. The problem with this theory is that there is no evidence for a gallon of this size ever having been used in England. Adams maintained, however, that the Irish wine gallon of 217.6 in^3, which was in legal effect until the adoption of Imperial measure in 1824, was a survival of this unit. If this little wine gallon actually existed in England, it must have been used in parallel with the Winchester gallon that is clearly defined by a plain English reading of the text and was embodied in Exchequer standards still in existence. Adams's interpretation does fit in very nicely with a theory regarding the origin of the English foot suggested in Chapter 6; as Adams points out, 216 in^3 is a cube $\frac{1}{2}$ foot on a side. Nicholson and other authors have also noted this, a point to which we will return further on.

Given the ambiguity of the Latin *galonem vini*, which can equally well be translated "gallon of wine" or "wine gallon," it's also possible that the Irish gallon was not copied from any physical standard but was constructed at a distance based on a misunderstanding of the *Tractatus*.[50]

Ancient lineage of tower weight

Unlike the troy pound, the pound sterling (tower pound) hasn't been used to weigh anything for the last five centuries, but it's worth noting that it, too, has ancient antecedents. The merchant's pound of 15 tower ounces is identical to a major ancient trade weight, the Attic mina (i.e., the mina of Athens, Greece). It is known that classical Greece was in regular trade contact with Britain (see note on page 39), and given the existence into medieval times of the Greek foot as a measure of length in English buildings, it's not hard to see how the Greek unit of weight came to be established there.

The Attic mina was 50 Attic Greek staters of 135 troy grains.[51] The half stater was called a drachma, and it weighed $67\frac{1}{2}$ grains.[52] The Attic drachma was further divided into six obols of $11\frac{1}{4}$ grains. The $22\frac{1}{2}$-grain sterling silver penny described in the *Tractatus*, therefore, was two Attic obols or $\frac{1}{3}$ of an Attic drachma. In fact, the Saxons made frequent use of a silver coin having the value of three silver pennies that was called a trimes or trims (after Latin *tremius*); this is simply the Attic drachma under another name.[53] As in all the ancient weight systems discussed here, the talent of the Attic tradition was 60 minas. Figure 4.4 shows the relationships between the classical Attic units and the medieval tower units.

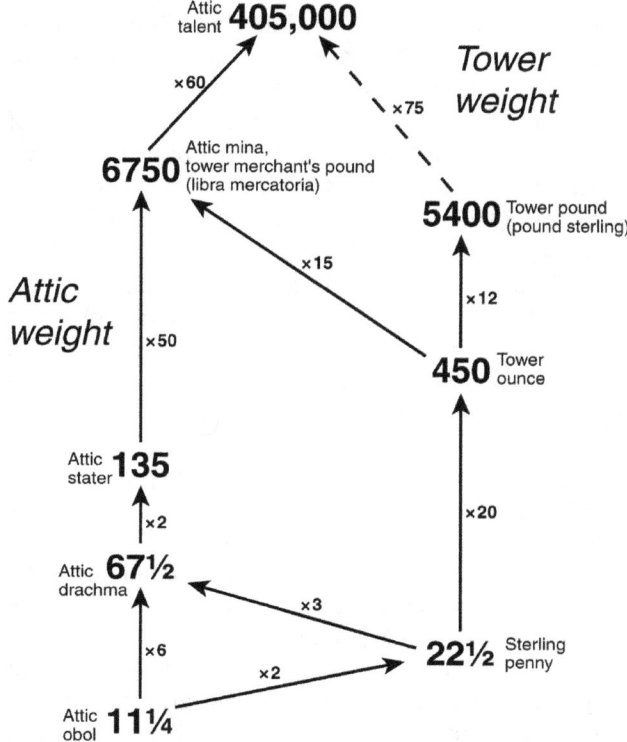

Figure 4.4: The ancient Attic and medieval tower weight systems

If you remember that the troy pound is the same as 40 ancient Egyptian qedets of 144 grains and that the troy pound also had a 15-ounce big brother, the pound of Cologne, the parallelism between Attic Greek (tower) weight and Egyptian (troy) weight is striking.

- 40 Attic staters = 1 tower pound
  = 12 tower ounces (5400 grains)
- 40 Egyptian qedets = 1 troy pound
  = 12 troy ounces (5760 grains)
- 50 Attic staters = 1 tower merchant's pound
  = 15 tower ounces (6750 grains)
- 50 Egyptian qedets = 1 pound of Cologne
  = 15 troy ounces (7200 grains)

The last two of these were the dominant premetric pound standards of southern Germany and north-ern Germany, respectively.[54] The four sets of relationships shown above can be seen as earlier and later manifestations of a single interrelated system with two "flavors," in earlier times illustrated by parallel Greek and Egyptian usages, then later by parallel German usages.

The pound of Cologne shows the unity underlying all this. As noted in Chapter 2, this basic Hanseatic trade unit was equivalent to 15 troy ounces, but its German users divided it into 16 ounces, each ounce being the same (in effect) as the tower ounce, 450 grains.[55] Coming at the same relationship from another direction, the mark of Vienna (the main standard for the Austro-Hungarian Empire, as noted above) was formally defined in relationship to the Pound of Cologne by a decree of Emperor Ferdinand I (1 August 1560) that set 5 marks of Vienna equal to 6 marks of Cologne;[56] if the mark of Cologne (half a pound) is 3600 troy grains, then the mark of Vienna is 4320 troy grains, which is 64 quentchen of $67\frac{1}{2}$ grains—the Attic drachma, as noted previously. (And 4320 is 10 of the units of 432 you saw in connection with troy weight, and it's $\frac{1}{100}$ of the ancient Syrian talent, and so on.)

Tower weight appears not to have been used in Roman Britain; the available evidence begins with the Saxon invasion of the 7th century A.D. The Saxon and Viking/Irish graves described in Chapter 2 that establish the existence of troy weight in England that far back also contain a smaller number of weights of the tower standard. Nine weights of this standard are based on a unit averaging 45.12 grains. These include a weight of 320 grains from a 6th or 7th century Saxon grave in Sarre, Kent, marked 7 (unit = 45.7 grains); a weight of 225 grains from a roughly contemporary Saxon grave in Gilton, Kent, marked 5 (unit = 45 grains); and another Saxon weight from Gilton weighing 180 grains and marked 8 (unit = $22\frac{1}{2}$ grains, the sterling penny).[57]

If tower weight came into England with the Saxons, then we would expect to find evidence at the source; and in fact the premetric unze of Dresden, the capital of the old Kingdom of Saxony, was 29.200875 grams or 450.6384 grains,[58] just 0.14 percent larger than the tower ounce in England.

A 1937 excavation at a Viking and pre-Viking site at Ronalsdway on the Isle of Man turned up

Chapter 4. The Winchester bushel

a bronze balance-beam and lead weight that show the tower standard in context and also show its use in parallel with troy weight.[59] A number of balances have been found in Saxon and Viking sites, but the design of the Ronaldsway balance is completely different, resembling instead a type common in Roman Britain. This type went out of general use after the 4th century, but Irish details date the find to the end of the 8th or early 9th century. The evidence suggests that the balance and weight were locally made on the island by the native Celtic inhabitants.

The weight can be used both as a pan weight and as a rider weight (like the weight that slides across the beam in older lab balances and the scales that used to be common in doctor's offices). It weighs 360.16 troy grains and is marked to indicate that it represents four units, each of which would therefore be 90.04 grains. This corresponds to half of a unit of 180.08 grains, eight units of 45.02 grains, or 16 sterling pennies of 22.51 grains, thus directly indicating a standard related to tower weight.

However, the balance beam itself has notches for the lead weight used as a rider, and in this mode, each notch represents a unit of about 15 grains. This division is not what would be expected if the balance had been intended to weigh Saxon pennies of $22\frac{1}{2}$ grains, but it is consistent with a system based directly or indirectly on the Attic tradition and Arabic coin weights of 45 and 450 grains (3×15 and 30×15, respectively). While most of the notches are very crudely spaced, giving increments ranging from 7 to 19 grains, the notches corresponding to 6 and 12 units are specially marked and are accurately located at positions of 90 and 180 grains, exactly—that is, four pennies and eight pennies, respectively, or one and two of the same 90-grain units shown by the rider weight itself. Even more interestingly, a third marked notch at 16 corresponds to a weight of 238 grains, which is almost exactly half of a *troy* ounce. This find suggests, therefore, that both tower and troy standards were in use in the same place; it cannot be explained in terms of Viking weights.

A find that does evidence the Viking weight system also shows a dual use with tower weight. A set of small ring-shaped gold weights discovered around 1924 in Rogaland (Norway) was found to have been cleverly designed to perform equally well against two different standards, one of 27.35 grams and one of 29.13 grams.[60] The first of these, at 422.1 grains, is the Viking eyrir (plural aurar), which is just the local version of the Roman ounce; the second, at 449.5 grains, is the tower ounce, or what in England became the tower ounce.

In the middle ages, the ancient Attic drachma of $67\frac{1}{2}$ grains persisted not just as the basis for the English tower pound (80 drachmas) and the libra mercatoria (100 drachmas) but also for several of the main coinage standards of the Continent as well.

Though not as widely attested as the troy cognates, a few units related to the ounce of 450 grains and the pound of 6750 survived in Arabic use into the 19th century. In Gedda, the derhem was 2.8835 grams or 44.45 grains;[61] contrast this with the troy cognate dirhem of about 48 grains found in most of the 19th century Arab world (Chapter 2). Another example from that region is Suakin, whose 19th century cossera of 60 rotolos had a value of 43.78422 kg; this can be interpreted as 1000 tower merchant's pounds (Attic minas) of 6756.938 grains, a scant 0.1 percent above the English version.[62] We can even find in 20th century Egypt a unit called the maqar of 3.51 grams or 54.17 grains,[63] which would be a decimal subdivision of the 5400-grain tower pound that parallels the 57.6-grain decimal subdivision of the troy pound.

In Paris, the pre-metric once for diamonds was 29.592 grams or 456.67 grains[64], clearly related to the tower ounce and the ounce of Cologne.

The most important 18th and 19th tower cognate is the common premetric weight of India, the tola. At its later British Imperial value of 180 troy grains, the tola is exactly $\frac{1}{30}$ of a tower pound or 8 sterling pennies; the native pre-Imperial values of the tola in Ferruckabad and Bengal (see Chapter 2) give equivalent tower pounds of 5407.42 and 5390.4 grains, respectively. To put it another way, the tola is what you get when you either divide the tower pound by 30 (as was the case with many of the ancient shekel/mina systems) or divide the troy pound by 32 (following the usual Germanic tradition).

In Calcutta, the sicca for gold, silver, coins, and precious stones was 11.642211 grams[65] or 179.67 grains; the tola ($\frac{5}{4}$ sicca) of 14.552764 grams was the same as 10 sterling pennies of 22.458 grains, about 0.2 percent lower than the English standard.

The old silver rupee of Bombay was 11.599 grams or 179 grains; 3000 of these rupees made the maund for spirits and arrack, at 34.797036 kg the equivalent of 100 tower pounds of 5370 grains.[66]

Weights and densities

Returning again to the relationships between capacity measures and weights and the parallel bases of both systems in the ancient shekel standards, we notice that certain proportions appear repeatedly throughout. Consider patterns based on a weight of 80 pounds, for example. In Chapter 3, I mentioned the well-attested equivalence of 80 Roman pounds of water or wine to the cubic Roman foot. We see something like this in our own early traditions, too:

- 80 tower pounds of wheat = 1 Winchester bushel (8 Winchester gallons)
- 80 troy pounds of wheat = 1 cubic Northern foot (10 wine gallons)

Another repeating pattern is the ratio $\frac{4}{5}$. Twelve ounces = $\frac{4}{5}$ of 15 ounces, so:

- tower pound = $\frac{4}{5}$ of tower merchant's pound (Attic mina)
- troy pound = $\frac{4}{5}$ of pound of Cologne (half sep, Syrian mina)

These common patterns arose from the need, noted earlier, to reduce all calculations of weight and capacity to their simplest terms. This applies not only to the relationship between units coming from different cultures but also to the relationship between the weights of different substances. In the latter case, the simplest terms are not arbitrary but are shaped by the physical characteristics of the trade goods themselves.

As you've already seen, the English systems were built around the physical properties of wheat and wine. Shippers handling these two most important ancient and medieval trade goods would need to equate the weight equivalents of their respective measures both to each other and to the the capacities of ships and storage vessels. It turns out, as a matter of empirical fact, that the simplest ratio between the weight of a given measure full of wheat and the same measure full of wine is $\frac{4}{5}$, or to put it another way, 40 to 50.[67] For example, if wine contains 12 percent alcohol by volume, then wheat that weighs $\frac{4}{5}$ as much will weigh 61.06 lb/bu[68] (see Appendix A for a discussion of grain weights to put this figure in context).

It follows, therefore, that

1 Winchester pint of wheat weighs
1 tower pound[69] (40 staters)

while

1 Winchester pint of wine weighs
1 tower merchant's pound (50 staters)

Similarly,

1 wine pint of wheat weighs
1 troy pound (40 qedets)

while

1 wine pint of wine weighs
1 pound of Cologne (50 qedets)

The grain next in importance to wheat was barley, and by some stroke of fate (or intelligent design, if you will), the simplest reasonable approximation to the ratio of the weight of barley to the weight of the same volume of wheat is the same as that between wheat and wine: $\frac{4}{5}$.[70] This ratio is reflected in our U.S. customary equivalents for grains measured by weight; a bushel by weight of barley is defined at the federal level and in most states as 48 avoirdupois pounds, while a bushel by weight of wheat is defined throughout the U.S. as 60 avoirdupois pounds.[71] The ratio $\frac{48}{60}$ is the same as $\frac{4}{5}$ or $\frac{40}{50}$. And this is why the number 80 seems to pop up so frequently; it's one of the few simple round numbers for a weight of wheat that yields equally simple round numbers for corresponding weights of barley and wine. Thus,

1 Winchester bushel contains
64 tower pounds of barley
80 tower pounds of wheat
100 tower pounds of wine

and

1 cubic Northern foot contains
64 troy pounds of barley
80 troy pounds of wheat
100 troy pounds of wine

These relationships ground our two most basic capacity measures—the Winchester bushel and the cubic Northern foot (10 wine gallons)—on the two most important ancient large weight units—the Syrian talent of 432,000 troy grains and the Aeginetan talent of 576,000 troy grains (see Chapter 2). If wine canonically weighs $\frac{5}{4}$ as much as wheat, then the grain weight equivalents stated in the old statutes mean that

   1 Winchester bushel of wheat
      weighs 1 Syrian talent

and

   1 cubic Northern foot of wine (10 wine gallons)
      weighs 1 Aeginetan talent

It's hard to imagine a simpler or more direct explanation of how these two basic capacity measures for wheat and wine originated.[72]

With the fundamental role of the $\frac{4}{5}$ ratio in mind, it's possible to assemble all the most important weight relationships discussed so far into a single synthetic overview, as shown in Figure 4.5.[73]

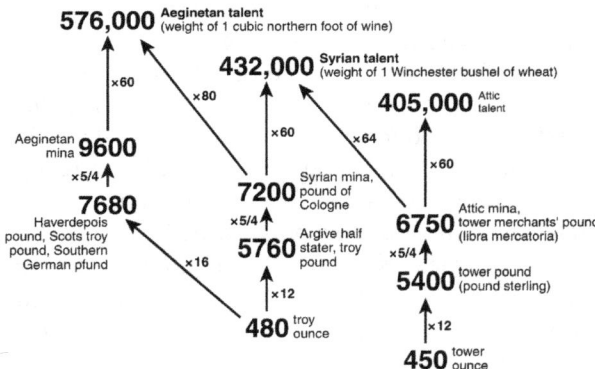

Figure 4.5: Synthetic overview of primary ancient and medieval standards of weight

The $\frac{4}{5}$ ratio between wheat and wine also allows us to extend the Northern foot basis of the wine gallon into an explanation of some Asian units. In Chapter 1, I pointed out that the Northern foot is found in Mohenjodaro (ca. 1500 B.C.) decimally divided into units of 1.320 inches that in China were called t'sun. And in Chapter 2, we traced the decimal division of the troy pound into a unit of 576 grains as far west as China and Japan, where it was called a tael or liang. If a cubic Northern foot of wine weighs 100 troy pounds or 576,000 grains, then a cubic t'sun weighs 576 grains—the Saxon Grove Ferry weight and (if we fill our cubic t'sun with water instead of wine) the slightly larger oriental tael.[74]

We can now also trace the Winchester standard back to an ancient original in terms of capacity. You'll recall from Chapter 3 the life-size paintings in the tomb of Hesy, an official of ancient Egypt, that showed two parallel series of capacity measures, one based on a unit of 28.8 in³ and the other on a unit of 21.6 in³. The former is the wine pint; the latter is $\frac{1}{100}$ of the Winchester bushel. This ancient unit, known as the Syrian/Phoenecian kotyle,[75] is the capacity measure that holds 1 tower pound of wine.

If you review the relationships we've covered in this chapter, you'll find that

   1 U.S. fluid ounce of wine weighs
      1 tower ounce

And this means that the kotyle, and the 21.6 in³ unit found in Hesy's tomb, are the same as our 12-ounce can or beer bottle. Next time you're holding a 12-ounce can of soda, note the weight; that's a tower pound.

Winchester measure as a Saxon import

As explained in Chapter 3, our other capacity standard—the one we use for liquids, the wine gallon—appears to go back in England to a time before the Roman occupation, and the same seems to be true of troy weight, with which it is closely associated. The tower pound comes in with the Saxons in the 7th century A.D., and it's closely associated with Winchester measure. It makes sense, then, to ask whether Winchester measure came in with the Saxons. If it did, we would expect to find traces of this standard in Germany, where it presumably originated.

Until the 20th century, most Germans measured grain by a unit called the scheffel. And of the four main scheffel standards—those of Prussia, Bremen,

Württemburg, and Dresden[76]—two are directly related to the Winchester bushel.

The Württemburg scheffel was 177.2263 liters,[77] which is 5.0293 Winchester bushels—that is, 5 Winchester bushels of 2163.002 in^3, about 0.6 percent above our value. You'll recall this unit from earlier in the chapter: it's the barrel for grain defined by the 1631 Virginia law requiring "that a barrell of corne shall be accounted five bushells of Winchester measure," a definition repeated in Virginia statutes five more times in the period from 1632 to 1661. The 1814 Virginia law defining a barrel of salt as 5 Winchester bushels shows that the scheffel of Württemburg was still part of the Saxon inheritance until relatively recently.

It's in Dresden, however, the capital of the old Kingdom of Saxony, that we find the original of the Winchester standard best preserved. There, until a reworking of the system in 1858, the scheffel was defined as 105.787583 liters,[78] which is 3.002 Winchester bushels—that is, 3 bushels of 2151.85 in^3, which is virtually identical to the Saxon standard as it's come down to us in Winchester measure.[79] The Dresden scheffel, just like its English counterpart, worked in a system of units strongly binary; so, very accurately,

 Scheffel of Saxony (4 viertel)
  = 3 Winchester bushels (24 gallons)

 Viertel of Saxony (4 metzen)
  = 6 Winchester gallons (24 quarts)

 Metze of Saxony (4 mässchen)
  = 6 Winchester quarts

 Mässchen of Saxony
  = 3 Winchester pints

Clearly these are two separate systems of dry capacity; just as clearly, they are carefully maintained descendants of a common ancestor—in this case, the cubic Greek cubit.[80]

It is reasonable, therefore, to conclude that both the tower pound and the Winchester standard of capacity were brought into England by the Saxons, and that our inheritance of two parallel weight/capacity systems came about from the introduction of the Saxon tower pound/Winchester system into a Celtic culture that already used a troy pound/wine gallon system, the older system probably having been introduced by Greek traders of the early classical era. This is not unlike the later Anglo-Saxon/Norman French mashup that gave us the English language.

However, the existence of these two families of capacity standards side by side in the ancient Egyptian tomb of Hesy (Chapter 3) and the evidence for native Celtic use of tower weight noted earlier in this chapter makes it at least possible that the two traditions existed together even before the Saxons entered England. But in the absence of evidence of the Winchester standard as old as the Carvoran measure, the thesis that Winchester came in with the Saxons remains the most likely.

The great confusion

To sum up our story so far: it appears from the historical data that there were two ancient English families of weight and capacity standards, a "Winchester family" based on tower silver coinage weights as discussed in this chapter and a "wine gallon family" based on troy gold weights as discussed in Chapter 3, and that both families were canonically defined in terms of theoretical wheat grains by roughly the same formula, the essential difference being that the weight units in one case were $\frac{15}{16}$ the size of the similarly named units in the other.

If you have been following the discussion very closely indeed, you will have noticed one remaining unexplained fact about the statute definitions of Winchester measure (this chapter) and the statute definitions of wine measure (Chapter 3). Once you understand the differences between the various pound standards, it's clear that the bushel defined on the basis of tower weight by the *Tractatus* (ca. 1301) is the Winchester bushel, and it's equally clear from actual weights of wheat that the gallon defined by Henry VII in 1496 in terms of troy weight (page 45) must be the wine gallon. The thing that's puzzled historians is the fact that the physical standards authorized by Henry VII in the act of 1496 "to remayne in his seid Tresory for ever" and still preserved in the Exchequer are undeniably Winchester standards, not wine gallon standards. This has led some to speculate that Henry's officials simply didn't know what they were doing, an assumption made more reasonable by the fact that the statute of 1496 was not Henry's first attempt to reform the state of English weights and measures. In 1494, Henry had authorized a complete overhaul of the standards, with results that are summarized in the preamble to the act of 1496 as follows:

> Where as afore this tyme the Kinge our Sovereign Lord intending the comen wele of his people, and to avoide the great disceite of Weightis and Mesures longe tyme used within this his Realme contrarie to thestatute of Magna Carta and othre estatutes thereof made by divers of his noble progenitours, att his great charge and coste did doo make weighits and mesures of brasse according to olde standardes thereof remaynyng in his Tresorye; and for that that oone weight and oone mesure shuld be used thrugh oute this his Realme, in avoiding of all fraude and discorde in that behalf, it was att the last parliament holden the xiiijth day of October in the xjth yere of our seid Sovereign Lordis reigne [1494] ordeyned that the seid mesures and weightis shuld be delyvered to the Knyghtes and Citezins of every Shire and Citie assembled in the same Parliament, Barons of the v. portes and certeyn Burgeises of Burgh Townes, surely by theym to be conveyed to certeyn Cities Burghs and Townes specified in a Cedule unto the same acte annexed there to remayne for ever, to thentent in the same acte more largely declared; whiche weightis and mesures upon more diligent examynacion had synz the making of the seid estatute been proved defective and not made according to the old lawes and statutes thereof ordeyned within the seid realm....

The king's displeasure over the waste of "his great charge and coste" comes through even in the language of the statute, and one can easily imagine Henry's trembling officials resolving to get it right the second time around.

The mystery of the missing wine gallon standards is solved by the existence of at least one set of city standards showing that Henry's workmen did construct and distribute official wine measures in parallel with the Winchester measures that are preserved in the Exchequer. A British government inquiry in 1909 found a set of four standards still in possession of the City of Bristol dating from the time of Henry VII and labeled A : WINE : POTTLE, A : WINE : QVART, A : WINE : PINTE, and A : WINE : HALFE : PINTE. All four are stamped with the initials H R (i.e., Henricus Rex) under a crown, and there can be no doubt that they were issued as part of the reforms that culminated in the statute of 1496 (the 1909 government report assigns the Bristol wine measures a date of 1495).[81]

To completely resolve this question, it's necessary to eliminate the possibility that these wine measures might actually be mislabeled versions of the Winchester measures in the Exchequer—in other words, Winchester standards being used to measure liquids, as seen in colonial laws cited earlier. In answer to a request I made in 1985, David J. Eveleigh, Curator of Agricultural and Social History at the City of Bristol Museum, very kindly provided me with measurements of these artifacts as shown in Table 4.1 (the two columns on the right are calculated from the measurements carried out in Bristol).

The average of the four implied gallons is 234.07 in^3. If we consider that smaller measures will always show greater errors in the implied gallon and exclude the half pint outlier, the average of the remaining three is 231.11 in^3.

The Bristol measures leave no doubt that the standards produced following the statutes of 1494 and 1496 included the wine gallon of 231 in^3 and its divisions and that the absence of wine gallon standards from the Exchequer, however it occurred, is not historically significant in this connection.

It appears, then, that the statute of 1496 captured the weight system of the wine measures, while the people in charge of making the actual standards for distribution thoughout the country provided both wine measures (for most liquids) and Winchester measures (for dry substances), preserving the two parallel systems that continue in the United States to this day.

The root of the debacle of 1494 is probably to be found in the reference to *Magna Carta* and Henry's assumption that it mandated "that oone weight and oone mesure shuld be used thrugh oute this his Realme." What the charter, of uncertain date but traditionally cited as 9 Henry 3 (1225), actually

Chapter 4. The Winchester bushel

Figure 4.6: "Trial of weights and measures under Henry VII." 1841 woodcut copy of an illustration from the time of Henry VII (now lost). The King himself supervises; the workman at top left checks troy weights, while the one at top right inspects a Winchester bushel; members of the Coopers' Guild see to the cask measures (defined in wine gallons); and standards found to be defective are thrown into the fire

NOTES

Inscription	Depth	Top inside diameter	Calculated volume	Equivalent gallon
WINE POTTLE	180 mm	115 mm	1869.64 cc	228.18 in^3
WINE QVART	131 mm	97 mm	968.07 cc	236.30 in^3
WINE PINTE	112 mm	73 mm	468.76 cc	228.85 in^3
WINE HALFE PINTE	88 mm	60 mm	248.81 cc	242.94 in^3

Table 4.1: Dimensions of the 1495 Bristol wine measures

required was "one measure of wine, one measure of ale, and one measure of wheat" throughout the realm—that is, three separate standards, one for each use. This appears to have been well understood early on; for example, a memorandum from the market at York dated 12 June 1393 records a purchase, authorized by the Mayor, of standards consisting of "a bushel, half bushel, and a peck," "a third of a gallon, gallon, pottle, and quart for ale," and "a gallon, pottle, and quart for wine."[82]

Despite the misunderstanding by a series of English monarchs of the intent of Magna Carta, separate wine gallon, ale gallon, and wheat gallon standards continued to exist in England until Imperial measure was instituted in 1824. If the 1495 date given the Bristol measures in the 1909 report is taken as exact, then it's probable that Henry, in attempting to reduce the measures to a single system, took the troy basis as definitionally correct and issued wine measures in 1495 (such as the ones still at Bristol) with the implication that they were to be used as grain measures as well; that the people, who traditionally used Winchester measure for that purpose, objected (anyone receiving rents of grain would have seen an immediate 14 percent reduction in income); and that the 1496 redistribution issued Winchester measures but kept the troy-based definition of the wine gallon. It's equally possible that by 1496 it was no longer understood that the system expressed in the *Tractatus* and the very similar system relating weight to the wine gallon actually referred to two different sets of weight standards, even though the physical measures themselves were very well understood to be distinct and were wisely carried forward as accurately as possible.

What is clear is that while the decision to continue both systems in parallel survives in active use in the U.S. more than 500 years later, the complex web of mathematical relationships underlying the two was never subsequently completely understood, and without a simple logical basis, the interlocking system became increasingly fragile. The weakness became fatal when the two parallel weight standards on which these systems were based were both replaced in popular usage with a third, unrelated weight standard—the avoirdupois pound.

Notes

[1] Zupko, *British Weights & Measures*, 11; SKINNER, 100. See also "Weights and Measures of the City of Winchester" at http://www.hampshire.gov.uk/regulatory/tradingstandards/wmhistory.html.

[2] As listed among the "commercial units of weight and measure in common use" by the U.S. Department of Commerce (*Federal Register*, Vol. 33, No. 146—Saturday, July 27, 1968, in Judson, *Weights and Measures Standards of the United States*, 35). Some early state laws rounded off 2150.42 to 2150.4. A 1798 Kentucky statute and an 1805 Ohio statute put the volume of the bushel at $2150\frac{2}{3}$ in^3 (ADAMS, 218; Littell, *Statute Law of Kentucky*, 2:194; *Statutes of Ohio and of the Northwestern Territory*, 519), and as late as 1897, the bushel was defined in Nebraska as 2150 in^3 (Brown and Wheeler, *Compiled Statutes of the State of Nebraska*, 1158). But these are trivial differences.

[3] Brigham, *Compact and Laws of the Colony of New Plymouth* (1836), 34.

[4] It is possible that this former bushel was the bushel of the wine gallon, used as it anciently could have been for dry measure as well as for liquids; but there is no physical evidence for such a unit.

[5] Brigham, *Compact and Laws of the Colony of New Plymouth*, 43.

[6] Lincoln, Johnson, and Northrup, *Colonial Laws of New York*, 1:64.

[7] Hening, *Statutes at Large... of the State of Virginia* (1809), 1:170.

[8] Ibid., 195.

[9] Ibid., 268.

[10] Ibid.

[11] Ibid., 472.

[12]Hening, *Statutes at Large ... of the State of Virginia* (1823), 2:90.

[13]Ibid., 196.

[14]Hening, *Statutes at Large ... of the State of Virginia* (1823), 3:254.

[15]*Code of Virginia* (1877), 487.

[16]Lincoln, Johnson, and Northrup, *Colonial Laws of New York*, 1:286 and 1:320.

[17]Phillips, *Revised Statutes of Louisiana*, 557.

[18]Letter Circular 1035, 26.

[19]In Wisconsin, it was changed to 100 dry quarts. To find this barrel defined in the North Carolina statutes, you have to know that 105 Winchester quarts is 3.28125 bushels:

> Whenever any commodity named in this section shall be quoted or sold by the bushel, the bushel shall be the number of pounds stated in this section and whenever quoted or sold in subdivisions of the bushel, the number of pounds shall be the fractional part of the number of pounds as set forth herein for the bushel, and when sold by the barrel shall consist of the number of pounds constituting 3.281 bushels (Chapter 81A–42, Standard Weights and Measures).

[20]With regard specifically to the vegetable and cranberry barrels, a 1912 report to the 62d U.S. Congress recommended

> That the standard barrel for fruits, vegetables, and other dry commodities shall be of the following dimensions when measured without distention of its parts: Diameter of head made of staves, seventeen and one-eighth inches; distance between heads, inside measurement, twenty-six inches; the outside bilge or circumference shall not be less than sixty-four inches; and the thickness of staves not greater than four-tenths of an inch: Provided, That any barrel of a different form having the same distance between heads and a capacity of seven thousand and fifty-six cubic inches shall be a standard barrel....
>
> The dimensions specified are those now established by law in many States, and are in customary and general use throughout the country.
>
> Provision is made that barrels used in packing or shipping cranberries may be of another size, because it has been demonstrated that these berries can not be packed without damage in the standard barrel (House Committee on Coinage, Weights, and Measures, *Standard Barrel for Dry Commodities*, 1–2).

It seems quite possible that a study of the barrels defined in older state laws in terms of their dimensions could yield further insights into the relationships underlying the Customary System of capacities; this avenue of inquiry remains open to anyone who would like to investigate it.

[21]Bruce, *Calendar of State Papers*, 24.

[22]*Calendar of Treasury Papers*, 2:467–87.

[23]Huckle, "A history of Beer and Pubs in Halesworth," referencing the *Manor of Halesworth Minute Book 1698*.

[24]Deetz and Fennel, *Records of the Colony of New Plymouth in New England*.

[25]The text here is the one published online by the Pennsylvania State Archives. A printed version can be found in Linn, *Charter of William Penn and Laws of the Province of Pennsylvania*, 115–16.

[26]*Notes and Queries* 11 S. III (Jan. 21, 1911) 56.

[27]But not the first statute mention of the Winchester bushel by name; that distinction belongs to 22 Charles 2 c. 8 §2 (1670).

[28]ADAMS (96–97) notes that a Massachusetts act of the year 1700 "prescribes, that the bushel used for the sale of meal, fruits, and other things, usually sold by heap, shall be not less than $18\frac{1}{2}$ inches within side; the half bushel not less than $13\frac{3}{4}$ inches; the peck not less than $10\frac{3}{4}$, and the half peck not less than 9 inches." It is probable that this series of diameters was traditional; as Adams notes, "The object of the provincial law was, to prohibit the use of bushels, which, though of the same cubical capacity, should be of shorter diameter and greater depth. It was for the benefit of the heap. It prescribed, therefore, only the diameter, without mentioning the depth; but that diameter, for the bushel, is identically the same, $18\frac{1}{2}$ inches, as the act of parliament... declares to be the width of the Winchester bushel in the exchequer."

[29]According to measurements carried out in 1931–32 (SKINNER, 105; CONNOR, 155), the Winchester bushel of Henry VII measures 2144.81 in^3 and the gallon, 268.43 in^3, corresponding to a bushel of 2147.44 in^3.

[30][Chisolm], *Seventh Annual Report*, 28.

[31]As measured in 1931–32 (SKINNER, 105; CONNOR, 160), the surviving Winchester standards of Elizabeth are: one bushel of 2148.28 in^3; one gallon of 270.59 in^3 (corresponding to a bushel of 2164.72 in^3); and another gallon of 268.97 in^3 (corresponding to a bushel of 2151.76 in^3). Other standard measures of Elizabeth belong to a series for the ale gallon of 282 in^3, which fell out of use in the U.S. in the 19th century.

[32]*Per discrecionem tocius Regni Anglie fuit mensura Domini Regis composita; videlicet, quod denarius Anglicanus, qui dicit^r Sterlingus, rotundus sit & sine tonsura, & ponderabit triginta & duo grana frumenti in medio spice; & viginti denā faciunt anciam; & duodecim uncie faciunt libram; & octo libre faciunt galonem vini; & octo galones faciunt bussellum London; quod est octava ps quartii.* The translation I've provided in the text is more literal than the one given in the *Statutes at Large*.

[33]This statute appears never to have been passed by Parliament but rather introduced "only by the supposed Royal Prerogative frequently assumed in that Reign" (*Report from the Committee* (1758), 19).

[34]The stone of $12\frac{1}{2}$ pounds, later displaced by the stone of 14 avoirdupois pounds still in use in England, has been largely ignored by historians, and there is scant physical evidence for it. That it existed, however, seems fairly certain; the Oxford English Dictionary has "1474 *stat. Winch.* in *Cov. Leet Bk.* 396 The wich kepes weyght & mesure l li. the halfe C, xxv^{ti} li. the quartern, xij li. & halfe the halfe quartern, þe wich was called of olde tyme beyng Stone of London, & vj li. & a quartern ys the halfe Stone, as it appereth in Magna Carta," i.e., 50 pounds in the half hundred, 25 in the quarter, $12\frac{1}{2}$ in the stone of London, and $6\frac{1}{4}$ in the half stone.

[35]*Per Ordinacionem tocius regni Anglie fuit mensura Domini Regis composita videlicet quod denarius qui vocatur sterlingus rotundus & sine tonsura ponderabit triginta duo grana frumenti in medio Spice. Et uncia ponderabit viginti denarios. Et duodecim uncie faciunt libram London. Et duodecim libre & dimid' faciunt petram London. Et octo libre frumenti faciunt galonem. Libra continet viginti solidos. Et octo galones faciunt bussellum London.* The translation given in the text is more literal than the one given in

NOTES

the *Statutes at Large.*

[36] To maintain consistency with earlier histories, I will follow tradition and continue to refer to this as the *Tractatus.*

[37] *Per ordinacionem Tocius Anglie regni, fuit mensura domini Regis composita, videlicet, quod denarius Anglicanus qui vocatur sterlyngus, rotundas & sine tonsura, ponderabit xxxij grana frumenti in medio spice. Et uncia debet ponderar̃ viginti denarios. Et duodecim uncie faciunt libram London. Et viii libre faciunt galonem vini. Et octo galones vini faciunt bussellum London. Et octo busselli faciunt quartium London. Et Duodecim libr̃ & di faciunt petram London.* The translation provided in the text is more literal than the one given in the *Statutes of the Realm.*

[38] SKINNER, 84–86. Skinner (88–89) suggests that the transmission of the 450-grain ounce and its addition to the 480-grain troy ounce already in use among the Saxons occurred when Offa, the Anglo-Saxon king of Mercia (A.D. 757–96) visited the court of Charlemagne, which had previously received Arabic weights, measures, and coins via emissaries from Caliph Harun-al-Rashid. According to Skinner, Offa began coining the first English silver pennies ($22\frac{1}{2}$ grains) in 791 following his return. According to Connor (109), the penny became standardized at $22\frac{1}{2}$ grains only after the Norman Invasion; but the existence in late antiquity of the Arabic silver dirhem at 45 grains (Hinz, *Islamische Masse und Gewichte,* p. 2, calculates the weight of the "classical silver dirhem" as 45.833 troy grains) makes it very probable that this was the origin of the English silver standard.

[39] A fact noted by Adam Smith in *The Wealth of Nations,* 1:39.

[40] *Rotuli Parliamentorum,* 5:109.

[41] *Item q̃ tout la moneie dor & dargent q̃ sera fait a la Tour de Loundres & a Caleis, ou aillours deinz le Roialme Dengleterre per auctorite roial, soit fait de auxi bone allaie & de joust pois come il est a la Tour au present fait.*

[42] *Report from the Committee* (1758), 20. See also ADAMS, 35; Henry, *History of Britain,* 251–52, citing Folkes, *Tables of English Silver Coins,* 1–2; [Chisolm], *Seventh Annual Report,* 16; Miller, "On the Construction of the New Imperial Standard Pound," 753–54.

[43] That is, according to "an official report," "less than the standard [troy] pound weight of the Exchequer by 15 dwt. 9 grs." ([Chisolm], *Seventh Annual Report,* 16). Nicholson (*Men and Measures,* 128) has this as 5404 grains, but as usual does not provide a source.

[44] Hall and Nicholas, *Tracts and Table Books,* 39.

[45] *Et sciend' quod quelibet libra de denariis & speciebus utpote in electuariis consistit solummodo ex pondere xx. s. Libra vero omnium aliarum rerum consistit ex viginti quinque solidis Uncia vero in electuariis consistit ex viginti denariis. Et libra continet xii. uncias. In aliis vero rebus libra continet quindecim uncias uncia est hinc inde in pondere viginti denariorium.* From the 18th century version of the *Statutes at Large.* The translation in the text is more literal than the one given in the *Statutes.*

[46] As I rely on Connor for many details, I must note that he disagrees (126–27) with the interpretation offered here, maintaining that the passage from the *Tractatus* defines two different pounds by referring to two different pennyweights and that one of these pounds (if I understand him correctly) is the troy pound. I will cite on the side of my interpretation Miller, "On the Construction of the New Imperial Standard Pound," 754–55; [Chisolm], *Seventh Annual Report,* 17; and SKINNER, 92.

[47] Henry, *History of Britain,* 257–58, quoting Brompton. A similar but not identical law of Æthelred's is cited in the online *Medieval Sourcebook* (http://www.fordham.edu/halsall/source/978ethelred-londonlaws.html) with a date of 978 A.D. I haven't seen the original of either of these laws. Henry (254) notes the opinion of some authors that the merchant's pound of 15 tower ounces was in use as far back as Æthelstan (first part of the 10th century), but there seems to have been some question about this.

[48] *Per ordinationem totius regni Anglie fuit mensura domini regis composita, videlicet, quod denarius Anglicanus qui vocatur sterlingus, rotundus et sine tonsura, ponderabit triginta duo grana frumenti, in medio spice. Et uncia debet ponderare viginti denarios. Et quindecim uncie faciunt libram Londonie. Et octo libre frumenti faciunt galonem vini, et octo galones frumenti faciunt bussellum London, hoc est, octavam partem quarterii. Duodecim libre et dimidia faciunt petram Londonie. Saccus lane debet ponderare viginti et octo petras....*

Sciendum est quod libra denariorum, specierum confectionum, utpote electuariorum, consistit in pondere viginti solidorum. Libra vero aliarum rerum ponderat viginti et quinque solidos. Item in electuariis, confectionibus, libra continet duodecim uncias. Et uncia consistit in pondere viginti denarios.... (Hall and Nicholas, *Tracts and Table Books,* 9.)

[49] Markham, *Records of the Borough of Northampton,* 1:327–28.

[50] I am indebted to my daughter, Clara Bosak-Schroeder, for pointing this out.

[51] CONNOR, 5; SKINNER, 60. The coin weight universally known as the stater should not be confused with the much larger earlier archaic stater mentioned in Chapter 2 as one probable basis for the troy standard.

[52] The canonical value of 270 grains for the tetradrachma, 135 for the stater, and $67\frac{1}{2}$ for the drachma appears to date to Barclay Head's 1887 analysis of Greek coins in the British Museum (*Historia Numorum,* xl-xliii.) An extensive 20th century statistical analysis of Athenian coinage puts the tetradrachma at 270.2 grains, which would make the drachma 67.55 grains (Hemmy, "Weight Standards of Ancient Greece and Persia," 81). In his summary of the weights (not coins) excavated at Olympia, Hitzl puts the drachma at 4.366 grams, i.e., 67.378 troy grains (*Gewichte griechischer Zeit aus Olympia,* 27). Head traces the 135-grain stater standard back to two 8th century B.C. Euboean towns, Chalcis and Eretrea. Corinth adopted the same standard, but divided the stater into three drachmas rather than two, putting the Corinthian drachma at 45 troy grains. The Athenians under Solon adopted the standard in coinage "distinguished by extreme purity of metal and by accuracy of weight, the full Euboïc weight of 270 grs. to the tetradrachm being more nearly maintained at Athens than anywhere else where the Euboïc standard prevailed" (*Historia Numorum,* xliii).

[53] The trimes was mistakenly called thrimsa or thrymsa by later authors; see the OED under "thrimsa." The drachma appears also to be the basis of the old standards of the Austro-Hungarian Empire. The premetric pfund of Vienna was 560.012 grams (TATE'S, 60; MARTINI, 827, has 560.06) or 8642.306 troy grains; this was divided into 32 loths of 270.072 grains (the Attic tetradrachma), and each loth was divided into 4 quentchens, which puts this Austro-Hungarian version of the drachma at 67.518 grains.

[54] SKINNER, 93. As noted in Chapter 2, Miller cites examples of 16 German city standards in the Hanseatic tradition averaging 7201.6 grains. Another group of 6 German city standards from the family represented in England by the tower pound av-

erages 6725.2 grains, which is about 0.4% lower than the unit the English ended up with (Miller, "On the Construction of the New Imperial Standard Pound," 755). A French document from the time of Edward III, "not long after 1329," puts the relationship between the English "Marc of Troyes" (i.e., 8 troy ounces) and the English "Marc de la Rochelle" (i.e., 8 tower ounces) at a proportion that places the tower ounce at 451.76 grains if the troy ounce is 480 (Miller, "On the Construction of the New Imperial Standard Pound," 754, citing Clarke, 15; Henry, *History of Britain* 252, citing Folkes, 90). Henry further points to the early 19th century ounces of Cologne and Strasburgh, both at 451.38 troy grains.

[55] Miller, "On the Construction of the New Imperial Standard Pound," 753; Henry, *History of Britain,* 252–53; SKINNER, 89.

[56] Kisch, *Scales and Weights,* 9.

[57] Smith, "Early Anglo-Saxon Weights," 122–29. The analysis here is mine, not Smith's.

[58] MARTINI, 199.

[59] Skinner and Bruce-Mitford, "A Celtic Balance-beam of the Christian Period."

[60] Bakka, "Two Aurar of Gold," 287.

[61] MARTINI, 222.

[62] MARTINI, 762.

[63] MARTINI, 58.

[64] MARTINI, 453.

[65] MARTINI, 130.

[66] MARTINI, 94.

[67] Adams appears to have been the first modern commentator to notice this.

[68] Wine that contains 12 percent alcohol by volume has a specific gravity (at 20 °C) of 0.98412 (*Encyclopedia of Industrial Chemical Analysis,* 19:406). Water in air at that temperature weighs 3774.653 grams per U.S. gallon or 252.4408 grains per cubic inch (*Handbook of Chemistry and Physics,* 60th Edition, F–2), so 12 percent wine weighs 248.4320 grains per in^3 and the rest can be calculated from various definitions already presented.

[69] And thus, of course, a Winchester quart of wheat weighs two tower pounds. This quart is probably the same as the old English measure known as the sester (after Latin sextarius), which we are told in a medical recipe from the early 11th century "shall weigh two pounds by silver weight" *(sceal wegan twa pund be sylfyrgewyht).* Löweneck, *Peri Didaxeon,* 11.

[70] See Appendix A for more detail.

[71] The weight equivalence for wheat, at least, appears to be fairly old; Adams, writing in 1821, notes that "the laws of many of the states require, that the bushel should contain 60 pounds avoirdupois of wheat."

[72] The old haverdepois pound (7680 grains) can be related to the Northern foot through another whole-number sequence for expressing barley–wheat–wine ratios: a cubic Northern foot contains 48 haverdepois pounds of barley, 64 haverdepois pounds of wheat, and 75 haverdepois pounds of wine.

[73] I should note here that this is not an overview of the entire complex of weight systems that came out of the bronze age and gave rise to most premetric systems of weight measurement. A complete account would include the binary/decimal weight system of Mohenjodaro and Dilmun, based on a shekel of 211 grains, which probably gave rise to Roman weights and their later Viking realizations, and a family of sexagesimal weights based on the Sumerian/Babylonian and later Persian standard called the Daric, after a gold coin of Darius weighing about 130 grains.

[74] And this cubic t'sun is equivalent to 6 cubic Roman digits. But I digress.

[75] Petrie, "Measures and Weights (Ancient)" 143; see also SKINNER, 43.

[76] *Myers Grosses Konversations-Lexikon,* 17:719. See also *Conversations-Lexikon Brockhaus,* 13:167.

[77] MARTINI, 749.

[78] MARTINI, 199.

[79] As noted earlier, the measure of 3 Winchester bushels remains the legal barrel in the state of Louisiana, though the earlier Louisiana statutes defining the barrel as $3\frac{1}{4}$ and then $3\frac{1}{2}$ Winchester bushels makes it doubtful that the unit comes from a Saxon original. This unit of 3 bushels has some interesting implications with regard to weight; 3 Winchester bushels of wheat weighs 240 tower pounds or 1,296,000 troy grains, which is 1 million Persian and earlier Sumerian shekels. (Several historians of the subject, including myself, put the canonical value of the Persian/Sumerian shekel at 129.6 troy grains as an aid to understanding its relationship to the other ancient systems we've been looking at. In other words, it's $\frac{9}{10}$ of an Egyptian qedet. I won't give sources here, as this is a subject for another book.)

[80] If the Greek foot is 12.44 inches, then the cubic Greek cubit is about 0.6 percent larger than the 19th century scheffel of Saxony.

[81] *Report by the Board of Trade on their Proceedings and Business under the Weights and Measures Acts, 1909,* 43.

[82] *Memorandum quod duodecimo die Junii 16 Ric. ii., deliberata fuerunt Willelmo Tundu, ex precepto maioris et consensu proborum, mensura et pondera subscripta, videlicet, unum busselum eris, dimidium bussellum et pek ligni, tercia pars lagene, lagena, potella, et quarta ligni pro cervisia, una trona ad ponderandum, j hoper cum scala, et j strykel ligni, lagena, potella et quarta eris pro vino. York Memorandum Book,* pt. 2, 13.

Chapter 5
The avoirdupois pound

As you learned in Chapter 2, the pound we use for weighing most things, called the avoirdupois pound, is not the same as the troy pound upon which weights for gold and silver are based. The troy pound of 12 troy ounces weighs 5760 grains, whereas the avoirdupois pound of 16 avoirdupois ounces weighs 7000 grains, the grain being the same in both cases.

Unlike all the other Customary units described so far—land measures, liquid capacities, dry capacities, and the troy units of weight—the avoirdupois pound was not inherited from the pre-Norman Saxons. The most convincing historical account of the avoirdupois pound traces it to a city standard of Florence, Italy,[1] and the connection isn't ancient, it's medieval, the standard coming into general use some time between 1250 and 1357.[2] Wool was in medieval times the chief export of Britain; in fact, a large square bag of wool, called the woolsack, is to this day the ceremonial seat of the Speaker of the British House of Lords, symbolizing the historic importance of wool to the British economy. And in medieval times, virtually all of the wool exported from Britain went to mills in Italy, where it was weighed using Italian standards.

In later centuries, the avoirdupois pound, probably because of its ubiquity as a wool weight in the larger sizes, eventually became the weight for all the bulky goods or "great wares" of trade, leaving troy weight for bread and precious substances or "subtle wares" weighed in small quantity, such as precious metals, medicines, and drugs.[3] Indeed, the word *avoirdupois* simply means "goods of weight," "having weight," or "weighable."[4] Wool was among the "things sold by weight" *(choses poisablez* in the words of a statute of 1429), as opposed to grains and liquids, which were sold by capacity measure.[5] The statute 9 Edward 3 stat. 1 (1335) refers to

Figure 5.1: *Avoirdupois 112-pound Exchequer standard of Elizabeth I*

"merchant strangers ... which do carry and bring in by sea or land, wines, avoir du pois, and other livings and victuals, with divers other things to be sold" *(marchantz estranges ... qi mesnent, carient, ou portent p meer & p tere vins, avoir du pois & autres vivres, vitailles, & autres choses vendables),* and the statute 27 Edward 3 stat. 2 c. 10 (1353) mandates that balances be used for weighing "all manner of avoir de pois" *(tut maner' de avoir de pois).*

The avoirdupois pound appears to have given rise to a gallon of 8 avoirdupois pounds of wheat by analogy with the wine gallon (8 troy pounds of wheat) and the Winchester gallon (8 tower merchant's pounds of wheat). This avoirdupois-based unit was the English ale gallon of 282 in^3.[6] By the $\frac{5}{4}$ ratio between the weights of wine and wheat noted

Chapter 5. The avoirdupois pound

in Chapter 4, the avoirdupois pound seems to have generated another gallon holding 8 pounds of wine. The standard of this gallon had a capacity of 224 in^3 and was kept at the Guildhall in London. It was not an official government standard but long served as the commercial standard for retail sale of wine in the city; the merchants paid taxes on barrels from France by the government wine gallon of 231 in^3 and then sold the same wine by a Guildhall gallon about 3 percent smaller, a practice that it took the government a couple of centuries to catch up with.[7] As noted in Chapter 3, it was the confusion between the Guildhall gallon and the wine gallon that led to the 1706 statute definition of the wine gallon we use today.

The avoirdupois system of weight

There are two systems of avoirdupois weight multiples going up from the pound, and one system of subdivisions going down. Let's take the subdivisions first.

	Unit	Pounds	Ounces	Drams
	pound	1	16	256
$\times \frac{1}{16} =$	ounce	$\frac{1}{16}$	1	16
$\times \frac{1}{16} =$	dram	$\frac{1}{256}$	$\frac{1}{16}$	1

The avoirdupois dram (sixteenth of an ounce) fell out of general use long ago.[8] It survives now only in the phrase "dram equivalent," used in connection with shotgun loads. The dram equivalent indicates the shot velocity imparted by a certain quantity of modern smokeless powder in terms of the weight of black powder that would formerly have been required to accomplish the same result.

The important relationship for understanding what goes on at the grocery store is just

1 avoirdupois pound = 16 avoirdupois ounces

This purely binary division is Germanic as opposed to the basically Roman 12-ounce division seen in the troy and tower weight systems, and it works very well down as far as the ounce. Below that, one begins to want to work in grains, but just as the overlay of two unrelated systems of length created extraordinarily clumsy definitions of the land measures in terms of feet, so the overlay of the unrelated avoirdupois measures upon the original troy basis created an extraordinarily clumsy and conflicting legal relationship between the two systems of weight at the level of troy grain equivalents. Varying 17th and 18th century estimates of the avoirdupois pound in troy grains[9] were finally resolved in the act creating Imperial measure by somewhat arbitrarily defining the avoirdupois pound as 7000 troy grains, leading to the stunningly ugly relationship

1 avoirdupois ounce = $437\frac{1}{2}$ troy grains

Just as with the land measures, the way to look at avoirdupois is purely on its own terms: ounces and pounds, forget the historically unrelated grains. But it must be admitted that to American eyes, the traditional units associated with avoirdupois look pretty strange, too.

	Unit	Pounds
	avoirdupois pound	1
$\times 7 =$	nail, clove, or half stone	7
$\times 2 =$	stone	14
$\times 2 =$	quarter or tod	28
$\times 2 =$	half hundred	56
$\times 2 =$	hundredweight (cwt)	112
$\times 20 =$	ton	2240

The stone is still used to some extent in the UK, and stone and clove standards appear together in the U.S. as late as a 1797 Vermont statute.[10] There was also a unit peculiar to the wool trade called the sack; it weighed 26 stone or 364 pounds.

A second system of multiples, common in the U.S., is

	Unit	Pounds
	avoirdupois pound	1
$\times 25 =$	quarter	25
$\times 4 =$	hundredweight (cwt)	100
$\times 20 =$	ton	2000

The quarter of this second series is no longer used as a named unit, though 25 pounds continues to be a common unit of sale.

Chapter 5. The avoirdupois pound

The word *avoirdupois*

In Chapter 2 you saw a brief history of the weight originally called haverdepois (Habur de Peyse, Haburdy Poyse, etc.), the pound of which (16 troy ounces) was also one of the two main German weight standards. Somewhere along the line, the big Germanic/troy haverdepois pound stopped being used, and the name became attached to the standard of the weights for wool; but exactly when this happened isn't clear. As mentioned earlier, the phrase *avoir du pois* originally just meant "goods of weight," and we don't find the word used to refer to a specific standard (the haverdepois pound of 7680 troy grains) until the late 1400s.

The distinctive 14-pound wool stone and 364-pound sack of 26 stone that belonged to the standard we now call avoirdupois go back much earlier in the statutes than any use of the name *avoirdupois* or its variants. 14 Edward 3 stat. 1 c. 21 (1340) and 15 Edward 3 stat. 3 c. 3 (1341) both state that there are 14 pounds to the stone for wool and 26 stone to the sack for wool, but neither mentions the word "haverdepois," and neither does 13 Richard 2 stat. 1 c. 9 (1389), which mandated "that none buy or sell wool at more weight than at fourteen pounds the stone" *(q̃ null homme achate ne vende leynes q̃ a quatorze livres le pere),* nor 11 Henry 7 c. 4 §2 (1494), which enacted "that ther be but only ... xiiij lb. to the stone of Wolle and xxvj stone to the sakke."[12] The clove or half stone appears earlier in 9 Henry 6 c. 8 (1430), which extended the wool weights to cheese by enacting "that the weight of a wey of cheese may contain xxxii cloves, that is to say, every clove vij li. by the said weights couching."[13] "Couching weights" or "lying weights" meant weights set against the goods in a balance, as opposed to the use of a weighing device called an auncel that was prohibited by the same statute. Again, there is no mention of any word like *avoirdupois* associated with the weighing of wool.

The earliest statute appearance of the name in connection with wool weights comes in 6 Henry 8 c. 9 (1514), which enacted

> that the Wolle which shalbe delyvered for or by the Clothier to any person or persons for breking kembing carding or spynnyng of the

Figure 5.2: Exchequer avoirdupois pound standard of Elizabeth I (bell form)

The relationship among larger weight units most important for Americans to remember is

1 ton (U.S.) = 2000 avoirdupois pounds

In the U.S., the British ton of 2240 pounds is called the long ton, and (if necessary to distinguish it) the ton of 2000 pounds is known as the short ton.

If you ignore the unwieldy troy grain equivalents, the American version of the traditional system is perfectly straightforward, and the use of binary going down from the pound and decimal going up provides a very handy set of units to use in practice.

Note the similarity of the American version of the series to the one defined in the *Tractatus de Ponderibus et Mensuris* (see pages 71, 74, and 83): stone of $12\frac{1}{2}$ pounds, quarter of 25, half hundredweight of 50, and hundredweight of 100. The standard of the *Tractatus* was the tower pound rather than the avoirdupois, but we seem to have inherited this system. Indeed, the oldest extant English standard weight of any kind is a lead shield weight of Edward I weighing 6.3 pounds avoirdupois, presumably intended to be $6\frac{1}{4}$ pounds;[11] this would fit the 100-pound series we use in the U.S. ($\times 2 = 12\frac{1}{2}$, $\times 2 = 25$, etc.) but not the 112-pound series.

Chapter 5. The avoirdupois pound

same the delyvery therof therfore shalbe by evyn juste and true poys and weight of haberdepoys sealid by auctoritie ...

This 1514 law perhaps marks the beginning of the confusion between the two uses of *avoirdupois,* but it must have taken most of the century for the name to become strongly associated with the pound we use now. 21 Henry 8 c. 12 (1529) specified that there be 20 pounds to the stone of hemp, extending the wool standard to weighing another fiber, but made no mention of avoirdupois. On the other hand, 24 Henry 8 c. 3 (1532) required beef, pork, mutton, and veal to be sold by "laufull weighte called Haberdepayes," but there is reason to believe that this referred to the pound of 16 troy ounces (Chapter 2), and no mention was made of cloves or stones.

If we go all the way back in the statutes, we find that the standard we call avoirdupois was not only not associated with that name, it was not even originally associated with the pounds we now call by that name; rather, it was associated only with the big wool weight units, the sack and the stone, and these were defined very roughly in terms of tower weight. The version of the *Tractatus* (traditionally cited as 31 Edward 1, 1301) given in the definitive *Statutes of the Realm* authorized by George III continues the passage quoted on page 71 as follows:

> A sack of wool ought to weigh 28 stones, that is, 350 pounds, and in some parts 30 stones, that is, 375 pounds, and they are the same according to the greater or lesser pound; six times 20 stones, that is, 1500 pounds, make a load of lead ... [14]

Recall that the pound specified in the *Tractatus* for heavier substances (like wool) is the tower merchant's pound of 15 tower ounces or 6750 troy grains (page 73). If the sack of 375 of these pounds is divided into 364 pounds according to the later statutes, the resulting pound for wool is 6953.98 grains, which is close to one of three values for the avoirdupois pound (in effect, 6960 grains) given in the London *Assise of Bread* book circa 1601. The oldest set of wool weight standards, dating to the time of Edward III (1340), contains weights of 91 pounds (the quarter sack), 56, 28, 14, and 7 pounds;[15] according to later definitions, the set implies an average avoirdupois pound of 6992 grains, but it's important to note that there appears to have been no actual pound standard at this time. This suggests that the sack was a standard originally independent of any of the English pound systems, as emphasized by the statement that it was "the same according to the greater or lesser pound," in other words, that its weight was independent of whether the pound of 12 ounces or the pound of 15 ounces was being used.

This line of reasoning is strengthened by early records of the wool trade. In the Southampton *Assize of Bread Book* for 1482–83[16] are recorded some two dozen transactions in wool, all of them stated in sacks and cloves, 52 cloves to the sack; there is no mention of pounds at all, or of stones, even when it becomes necessary to specify weights using fractions of the larger units.

A manuscript from the 1400s intended as instruction for the beginning trader lists all the wool weights, including one I haven't mentioned till now (the sarpler, equal to two sacks). The writer gives the later pound equivalent for each—claw (clove) or nail 7 pounds, stone 14, tod 28, sack 364, and sarpler 728—and then continues

> Butt as fore Powndes and Sarplers, thai be butt lityll usyd in beyng and sellyng among merchauntes; for thai use to by or sell most comynly odyr by the Clawe, the Nayle, or by the Stone or the Todde, or the Sake.[17]

The sequence of events that emerges from these considerations looks something like this:

– The sack for wool is adopted at a value close to its traditional weight by 1301 and roughly specified for statute purposes in terms of tower merchant's pounds

– By 1340, the sack is divided into 26 stone and the stone into 14 wool pounds of about 7000 troy grains, probably to align with the Italian pounds used on the other end of the trading relationship; but a century later, the wool pounds are still not commonly used, nor is the name *avoirdupois* commonly attached to them

– Somewhere between 1495 and 1514, the name *avoirdupois* begins to end its association with the haverdepois pound of 16 troy ounces and starts attaching to the wool pounds, which come to be used for weighing most bulk goods, a process that takes the rest of the century to complete.

The downfall of the bushel weight systems

The existence of two similar families of weight and capacity standards, a "silver/Winchester bushel system" based on the 450-grain tower ounce and a "gold/wine gallon system" based on the 480-grain troy ounce, must have created confusion as early as the 1400s. The introduction of the wool pound into this fragile set of relationships finally destroyed it, as evidenced by the laughably inaccurate equivalences between the different pound units in various late medieval accounts of British weights.

The 1574 report of the jury appointed by Elizabeth I to investigate the state of the Exchequer standard weights shows a point at which the confusion had become permanent. These worthies, who included the Lord High Treasurer and the Chancellor of the Exchequer as well as an assortment of leading London aldermen, merchants, and goldsmiths, reported back as follows:

> to the first of the same articles theye saie that ther are onlie twoo sortes of weights lawfull in use at this daie with in this realme of England, and none other so fare as to them doth in any wise apeare. To the second article they saye, that the one sorte of weight nowe in use is commonlie called the troie weight, and the other sorte thereof is also comonlie called the avoir de poiz wieght, and further they say that both the saide consiste compounded from thauncient Englishe penye named a sterling, rounde and unclipped, which penny is limeted to waie twoo and thirtie graines of wheate in the midest of the eare, and twentie of those pence make an oz., and twelf of those ounc make one pounde troie. To the thirde and fourthe articles they saie that the said twoo sortes of weights doe differ in weight the one from the other three ounces troie at the pounde weight, for the pounde weight troie doth consiste onlie of xii oz. troie, and the lb. weight of avoir de poiz weight dothe consiste of fiftene ounc troie, and they saie that according to the auncient usage and longe custome of time, wherof no memorie is to the contrary, the troie weight is to be used in the weainge of breade, gold, silver, pretious stonnes, pearles, corall, amber, and kindes of confections namelie, electuaries. And y^t in all other thing the avoir de poiz weight is to be used.[18]

Here the ancient system was dutifully (and flawlessly) recited by the jury; but alas, the old troy merchant's pound (pound of Cologne) of 7200 grains (15 troy ounces) had become fatally confused with the wool weight of about 7000 grains that by this time had become the merchant's pound in practice. Given this misunderstanding, the jury could hardly find otherwise than

> that the weights to them delivered by order of this corte and all other by them examined are uncertaine and not of the right standard of England....

The confusion was not just with the 15-troy-ounce pound of Cologne but also with the 16-troy-ounce haverdepois pound as well. Just three years after the report of 1574 quoted above, a comprehensive treatise on the history and customs of the British isles, Holinshed's *Chronicles of England, Scotland, and Irelande,* stated that

> We have also a weight called the pounde, whereof are two sortes the one taking name of Troy contayning twelue ounces (after which our liquide & drie measures are weighed and our plat [silver] solde) the other commonly called *Haberdupois,* whereby our other artificers and chapmen doe buye and sell theyr wares. The first of these contayneth *7680.* graines [wheat grains, 32 to the penny per the old statutes; i.e., 5760 troy grains] wheras the other hath *10240* [i.e., 7680 troy grains]....
>
> Hitherto also I have spoken of small weightes, nowe let us see what they be that are of the greater sort, but first of such as are in use in Englande, reconing not after troye weight,

Chapter 5. The avoirdupois pound

but Haberdupois, whose pound hath sixteene ounces as I have sayde before. Of great waight therefore we have

The cloue weighing *7* pound or half a stone.

The halfe quarterne of *14.* pounde, in wooll a stone, whereof *26.* do make a sacke.

The quarterne of *28.* pound, in wool a Todde.

The halfe hundred of *56.* pounde.

The hundred of *112.* or *1792.* ounces.[19]

Notwithstanding the authoritative definition of the haverdepois pound as 16 troy ounces (7680 troy grains) in this passage and the equally authoritative definition of the same pound as 15 troy ounces (7200 troy grains) in the preceding one, we know from the surviving artifacts that the clove, stone, sack, and tod used for wool were actually the specified multiples of the (roughly) 7000-grain pound we now call avoirdupois. It's clear from these examples, therefore, that the old definitions and customary relationships continued in force on paper but had stopped bearing any relationship to the weights actually used in the marketplace for "gross wares."

In the midst of this confusion, it's not surprising that the reworked standards attempted by Elizabeth's jury of 1574 failed to meet with popular assent. A second jury consisting of 18 merchants and 11 goldsmiths of London, empaneled in 1583, took the easiest and probably wisest course, gathering together the best representative samples of the troy weight they could find, making fresh troy Exchequer standards from those, and creating a new avoirdupois standard by taking the oldest representative Exchequer wool weight then still in existence (a 4 stone or half hundred weight of Edward III) and dividing that to create the avoirdupois pound and its multiples. In this they acted much as Henry VII's commissioners appear to have done for the bushel standard in 1496 (Chapter 4). The second effort met everyone's approval, and the Exchequer weights created by the jury of 1583 set the troy and avoirdupois standards we use today.

It's important in connection with the apparently late division of the wool weight into pounds noted above that the standard avoirdupois pound of Elizabeth was not based on previously existing pound weights (which were considered too variable) but

Figure 5.3: Exchequer avoirdupois pound standard of Elizabeth I (disc form)

rather on the half hundred wool weight of Edward III, which was the oldest Exchequer example of the standard still existing (it subsequently went missing and by the time of the Committee of 1758 could not be found).[20] So in this sense it can be said that the avoirdupois pound as it has come down to us is an Elizabethan creation.

With the institution of "separate but equal" troy and avoirdupois Exchequer pounds, the two standards staked out mutually exclusive areas of application in which there would never be occasion to accurately compare the two, thus allowing them to maintain their respective values despite varying customary definitions of their supposed relationship. But the use of avoirdupois for weighing grain in addition to its original role for weighing wool took away the original tower weight basis of our bushel.

The good news: pounds and gallons

Fortunately for us, however, the unit we ended up with for weight—the avoirdupois pound—and the unit we ended up with for liquid capacity—the wine gallon—are exactly related by weight of water in a very useful way. The relationship is

12 wine gallons of water weighs 100 avoirdupois pounds

In other words,

A dozen gallons of water weighs a hundredweight (cwt)

This is not merely close; it's an exact equivalence for distilled water at about 51.2 degrees Fahrenheit.[21] In practical terms, it's even better than the original relationship between the liter and the kilogram; that definition (since abandoned) was based on the weight of water near freezing in a vacuum, whereas this is for water weighed in air, where most of us live, and at a temperature that's typical for water as it comes out of the ground.

The basis on a dozen makes a lovely set of relationships for goods as they are actually sold: a six-pack of U.S. pints contains 100 avoirdupois ounces of liquid by weight, and a six-pack of 12-oz. beverage cans contains 75. Again, these equivalences are exact for water weighed in air. A case of pints holds three gallons—the same as the Carvoran measure described in Chapter 3—and contains 25 pounds (a U.S. avoirdupois quarter) of liquid.

It's hard to tell, however, whether the wine gallon and the avoirdupois pound are actually historically related. As previously indicated, all the ancient units were related as simply as possible to each other through weights of basic trade substances,[22] and it therefore follows that their derivatives, if accurately kept, bear simple implied ratios to each other—even if there is no actual historical connection but only an indirect mathematical connection.

Unlike the abundant evidence for troy weight in England going back to early Saxon times, there appears to be no evidence for avoirdupois weight in England before the medieval wool trade; so if we accept that the relationship is too exact to be coincidence, then it is either the result of an undocumented medieval adjustment to align a historically unrelated Italian standard with existing capacity measures—which, as I will argue further on in a related context, is certainly possible—or it is just a mathematical consequence of the relationships already built into the ancient systems.

Whichever way it happened, the relationship between avoirdupois pound and wine gallon is a very useful one.

The decimal pound

As noted before, the 1824 definition of the avoirdupois pound as 7000 troy grains makes for an extraordinarily clumsy definition of the avoirdupois ounce: $437\frac{1}{2}$ troy grains. In fairness, however, it must be conceded that it's not a bad fit for a pound that's divided up decimally rather than into ounces, because it allows the pound to be divided up into tenths, hundredths, and thousandths in whole numbers of troy grains (700, 70, and 7). Indeed, the commission formed to restore the standards of weight and measure after the fire that destroyed the houses of Parliament, which argued strongly for the complete decimalization of the system, recommended in its 1841 report "that the name *millet*, or some other to be fixed by Act of Parliament, be recognised as describing the thousandth part of the pound, without the necessity of further definition."[23] Whether the commissioners were aware that millet (the seed) served as the traditional basis for Chinese weights is an open question; it would seem to be quite the joke for such a dour group. Their recommendation that a special name be created for the unit of 7 troy grains (a thousandth of a pound) went nowhere at the time, but a visit to the deli or bulk foods department of any modern U.S. supermarket will show that they won in the end: all electronic grocery scales nowadays read out in hundredths or thousandths of a pound and don't display ounces at all.

Notes

[1] CONNOR, 132; also SKINNER, 97. The 19th century value of the libbra in Florence (MARTINI, 207) was 339.542 gm or 5239.93 grains; this is 12 oncia of 436.66 grains, about 0.19 percent smaller than our U.S. avoirdupois ounce of $437\frac{1}{2}$.

[2] CONNOR, 129.

[3] The distinction between "great wares" (or "gross wares") and "subtle wares" served also to distinguish heavy goods weighed by the hundredweight of 112 pounds (always associated with avoirdupois) from fine goods weighed by the hundredweight of 100 pounds, as in a passage from the 15th-century MS Cotton Vespasion E IX ff. 86–110 cited by Connor (140). The OED gives the meaning of *subtle* (or *subtile*) applied to weight as "of weight, after tare has been deducted," and in some cases cited (especially when the word is spelled *suttle*) this appears to be true. But in other cases, including possibly some cited by the OED, the word distinguishes one weight standard from another, in particular, troy weight from avoirdupois. In the Ionian

NOTES

Islands, for example, which used British units well into the 20th century but gave them Italian names, the troy pound was known as the libra sottile and the avoirdupois pound as the libra grossa (TATE'S, 100).

[4]The word *avoir* can mean either "having" or "goods, property," the second corresponding to the obsolete English word *aver;* they all come from Latin *habere,* to have.

[5]ADAMS, 37.

[6]If a wine gallon of 231 in^3 holds 8 troy pounds of wheat, as stated in the statute of 1496, then 8 avoirdupois pounds of that wheat will occupy 280.73 in^3. The Exchequer ale quart of Elizabeth I measures 70.22 in^3, which corresponds to a gallon of 280.88 in^3 (CONNOR, 160). I cannot now recall whether I was the first to make this observation; if I'm not, I don't remember the source.

[7]ADAMS, 41.

[8]The Parliamentary commission of 1841 found that "the avoirdupois dram ... does not appear to be used at all" (*Report of the Commissioners, 1841,* 8), though there were standard weights of 8, 4, 2, 1, and $\frac{1}{2}$ dram still in the Exchequer (ibid., 10). In fact, weights expressed in a combination of avoirdupois pounds, ounces, and drams can be found in print as late as the 1955 UN Handbook referenced in Chapter 2.

[9]Uncertainty about the equivalence of troy and avoirdupois weight persisted until surprisingly late. Typical of the less inaccurate estimates is the 1819 Illinois statute that defined the avoirdupois pound as "seven thousand and twenty grains, troy or gold weight" (*Revised Code of Laws of Illinois,* 190). The Committee report of 1759 (p. 9) put the avoirdupois pound at 1 pound 2 ounces 12 pennyweights troy or 7008 grains, the same figure found in a passage from the *Assise of Bread* (1601) that equated the avoirdupois pound with $14\frac{1}{8}$ ounces and 2 pennyweights troy; but the same *Assise of Bread* equated 7 pounds avoirdupois with 102 ounces troy, which would make the avoirdupois pound weigh 6994 grains, and it also equated 14 pounds avoirdupois with 16 pounds 11 ounces troy, which would put the avoirdupois pound at 6960 grains.

[10]The treasurer of the state was required to keep standard weights of 56, 28, 14, 7, 4, 2, and 1 pound avoirdupois (*Laws of the State of Vermont,* 400). An 1807 law establishing weights and measures for the Mississippi Territory (later the states of Mississippi and Alabama) required a 14 pound weight among the standards to be maintained in each county (*Statutes of the Mississippi Territory,* 428; Aiken, *Digest of the Laws of the State of Alabama,* 445), and a similar 1803 Tennessee law required a 7 pound weight in the county standards (Scott, *Laws of the State of Tennessee,* 1:775).

[11]CONNOR, 233.

[12]The OED (s.v. "Sack") quotes the statute as saying "xxj stone to the sakke"; this is a misprint.

[13]$\bar{q}$ *le poys dune Waye dune formage* [sic] *puisse tenir xxxij cloves, cetassavoir chacun clove vij li. p les ditz poisez chochantz.*

[14]... *Et Duodecim libr̃ & di faciunt petram Londoñ.* ¶ *Saccus lane debet ponderare xxviij petr̃,* ([interlined] *hoc est CCC. & l. li.*) *Et in aliquibus partibus xxx petr̃* ([interlined] *hoc est CCClxxv. li.*) *Et idem sun sc̃dm majorem & minorem libram.* ¶ *Sexies viginti petre* ([interlined] *hoc est xv C. li.*) *faciunt charr̃ plumbi...*

[15]SKINNER, 96–97.

[16]13–17.

[17]Hall and Nicholas, *Tracts and Table Books,* 19.

[18]Harl. MS 698, f. 104; quoted in [Chisolm], *Seventh Annual Report,* 12.

[19]Holinshed's *Chronicles,* vol. 1, bk. 3, 121. This is the first edition, not the revised 1587 edition that served as a source for Shakespeare's histories but omitted the section on weights and measures.

[20]*Report from the Committee* (1758), 52–53, 61, 64. See also *Archeological Journal* 30 (1873): 430.

[21]See the table "Weight of one gallon of water (U.S. gallons)" in the *Handbook of Chemistry and Physics.* The version in the 60th edition (1979) gives the weight of a gallon of water at 10 °C as 8.33383 pounds avoirdupois and the weight at 11 °C as 8.33307 pounds avoirdupois; this puts the gallon of $8\frac{1}{3}$ pounds at 10.65 °C or about 51.2 °F.

[22]I believe that in the oldest systems, the two primary substances were barley and water, between which the simplest ratio is 5 to 8.

[23]*Report of the Commissioners, 1841,* 9.

Chapter 6

King Henry's arm

The English standard for shorter length measures (the inch, foot, and yard) is unusual among major premetric length standards of Europe in having no clearly identifiable ancient ancestor—neither the Roman foot of 11.64 inches nor the Greek foot of 12.44 inches nor the Northern foot of 13.2 inches.

The English *system* is unremarkable; the division of the foot into twelve units (notionally the width of the thumb at the root of the nail) was common from Roman times and is seen in most pre-metric European systems of length. This unit was usually called by the local word meaning "thumb." Many features of our traditional system are common to other European systems as well, including the tendency to give length units mnemonic names based on body parts. Table 6.1 lists the length units of the Customary tradition.[1]

Many of these units or their equivalents, most of them rarely used nowadays in the Customary system, were common in premetric European systems of length measurement regardless of the specific standard in terms of which they were defined. But the particular English standard that survives in the U.S. Customary System, the yard whose metric equivalent is 0.9144 meters, is an early medieval invention, first appearing in the 10th century A.D. and defined by law in terms of the older land measures (through the relationship $5\frac{1}{2}$ yards = 1 rod) by the end of the 12th.[2]

Once established, the English yard standard was maintained with exquisite care. The oldest surviving example is the silver yard of the Merchant Taylors' Company, which dates from 1445 and measures 36.001 inches of our current standard; the next oldest is the Winchester City Yard of 1497, which measures 35.996 inches.[3] Thus the standard has remained unchanged for more than 800 years; but its origin is a historical mystery.

Unit	Yards	Feet	Inches	Digits
fathom	2	6	72	96
geometric pace	$1\frac{2}{3}$	5	60	80
ell	$1\frac{1}{4}$	$3\frac{3}{4}$	45	60
yard	1	3	36	48
military pace	$\frac{5}{6}$	$2\frac{1}{2}$	30	40
cubit	$\frac{1}{2}$	$1\frac{1}{2}$	18	24
foot	$\frac{1}{3}$	1	12	16
span	$\frac{1}{4}$	$\frac{3}{4}$	9	12
shaftment	$\frac{1}{6}$	$\frac{1}{2}$	6	8
hand	$\frac{1}{9}$	$\frac{1}{3}$	4	$5\frac{1}{3}$
palm	$\frac{1}{12}$	$\frac{1}{4}$	3	4
nail	$\frac{1}{16}$	$\frac{3}{16}$	$2\frac{1}{4}$	3
inch (thumb)	$\frac{1}{36}$	$\frac{1}{12}$	1	$\frac{4}{3}$
digit (finger)	$\frac{1}{48}$	$\frac{1}{16}$	$\frac{3}{4}$	1

Table 6.1: Traditional units of length

An account of the life of King Henry I written by William of Malmesbury (1095–ca. 1143) states that "the measure of his own arm was applied to correct the false ell [yard] of the traders and enjoined on all throughout England."[4] The timing is about right (Henry I lived 1068–1135), but it's highly unlikely that a national standard was set so arbitrarily; that simply doesn't fit the universal ancient respect for standards and the almost universal care with which they were maintained, the occasional lapse in ability notwithstanding. The naive belief that a major standard would be defined this way is not far from the Victorian conception that the English yard was "founded upon the breadth of the chest of the Saxon race."[5] It's more likely that the standard was chosen by experts in the relevant trade guilds for practical

Chapter 6. King Henry's arm

reasons beyond the understanding of the average medieval English peasant and that it was easiest to explain the length arrived at as an exercise in pure monarchal authority.

I don't think any real scholar of the subject believes the hokum about King Henry's arm, but that doesn't mean that anyone has a good alternative explanation, just a suspicion that there's more here than meets the eye. Skinner notes that the English foot was "evidently a compromise between competing interests";[6] and Connor says, "One can but speculate how the present twelve-inch foot came to be adopted. Perhaps other interests had to be served."[7]

In what follows, I'll review some empirical facts regarding the relationship between the later length standard and the framework of weight and capacity measures in use at the time that may shed some light on the "other interests" that led to its adoption. Along the way, I'll present four more or less independent theories for the definition of the modern inch/foot/yard standard; the truth may involve any or all of them.

Theory 1: roundoff error

The simplest—and to my mind, the least persuasive—theory for the Customary foot and yard is that the statute setting $5\frac{1}{2}$ yards equal to a (Northern) rod was originally a rough equivalence, not an exact one.

As you learned in Chapter 1, many Anglo-Saxon structures were laid out using the Northern rod (our present Customary rod of $5\frac{1}{2}$ English yards). For purposes of discussion, let's use the figure that the archaeologists cite: 5.03 meters. But another rod of 4.65 m is also found, sometimes in the same place. The rod of 4.65 m survived into the 19th century in the Saxon homeland in Germany, suggesting that this is the unit brought over by the Saxons in the 7th century and that the 5.03 m Northern rod is a native (Celtic) unit.[8] The 4.65 m Saxon rod would have been based not on the later English yard but on a Saxon yard ($\frac{1}{5}$ rod) of about 36.6 inches. If one were trying to express the length of this implied Saxon yard in terms of the Northern rod, one would get

Northern rod = 5.41 Saxon yards

and then (the theory goes) if one wished to express this relationship very simply, one would round off to get

Northern rod = $5\frac{1}{2}$ yards

as in the statute, thus transforming an imported Saxon foot of about 12.2 inches into the later English foot as the result of a roundoff error.[9]

This theory is unsatisfying in light of the fact that the native Northern unit eventually won out over the Saxon import, which had vanished from use by the time the statute was enacted. It doesn't make sense that the statute would freeze into law an incorrect relationship with a standard of length that had already disappeared. Rather, the law was intended to carry the English yard standard forward by defining it in terms of the ancient land measure.

Given the fact that there is no archeological evidence for the use of the English yard before the Norman invasion, we need to find a reason for its creation.

Theory 2: densities and cubic inches

In Chapter 4, we looked at the barley–wheat–wine weight (density) ratios that provide an explanation for the repeated appearance of the number 80 in connection with various cubic foot standards (page 77ff.). The same density ratios also provide one explanation for the emergence of the inch as a unit of convenience in dealing with weights of these basic trade goods in an environment that already had to juggle two existing sets of capacity measures, one based on the wine gallon and one based on the Winchester bushel.

Consider the following:[10]

If one wine gallon (our present U.S. gallon) of wheat weighs 8 troy pounds, as stated in the statute of 1496, then a cubic inch of wheat weighs 199.48 grains, and the 64:80:100 proportion between the weights of canonical wheat, barley, and wine gives us the following set of equivalences:

Chapter 6. King Henry's arm

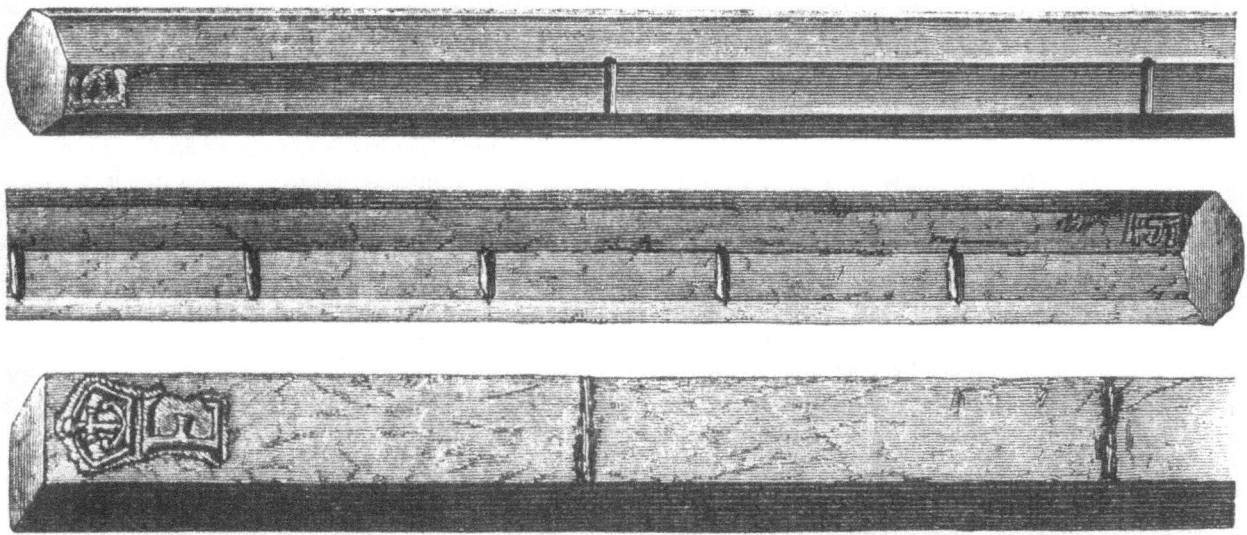

Figure 6.1: Exchequer standard yards. From top to bottom: standard yard of Henry VII, marked in nails (sixteenths of a yard); same on other end, marked in inches; standard yard of Elizabeth I, marked in nails

barley weighs	159.58	grains per cubic inch
wheat weighs	199.48	grains per cubic inch
wine weighs	249.35	grains per cubic inch

On the other hand, if one Winchester bushel (our present U.S. bushel) of wheat weighs 80 tower pounds, then

barley weighs	160.71	grains per cubic inch
wheat weighs	200.89	grains per cubic inch
wine weighs	251.11	grains per cubic inch

Split the difference and you get

barley weighs	160	grains per cubic inch
wheat weighs	200	grains per cubic inch
wine weighs	250	grains per cubic inch

All of which are excellent working approximations to the relative weights of real barley, wheat, and wine.[11] Note that this is the same 64–80–100 series we looked at in Chapter 4 multiplied by $2\frac{1}{2}$.

If this last set of relationships is taken as exact, then it follows that

> 80 tower pounds of wine occupy 1 cubic English foot

In other words, the cubic English foot is the measure that holds the quantity of wine whose weight is equal to the weight of wheat in the Winchester bushel.[12]

The 80 tower pounds of wine in the cubic English foot mirrors the 80 troy pounds of wheat in the cubic Northern foot and the 80 Roman pounds of wine or water in a cubic Roman foot. If the length of the English foot is taken as given, then it follows that a Winchester bushel should hold 2160 in^3, that is, $\frac{5}{4}$ of a cubic foot (1728 in^3); this is only half a percent larger than the measure we actually ended up with.[13] For all practical purposes, a U.S. bushel and $1\frac{1}{4}$ cubic feet can be taken as the same thing, so that, for example, 8 bushels is practically the same as 10 cubic feet, and a cubical granary 10 feet on a side holds about 800 bushels. I am, of course, not the first person to have noticed this; for example, an American textbook on business methods from 1911 simply says "1 cubic foot, four-fifths of a bushel."[14]

If the cubic English foot developed from Winchester measure, it's not hard to see why the linear English foot couldn't be made to fit the linear Northern foot in any sensible way; the standards are related by their volumes, and in comparing the lengths, we're looking at the relationships between their cube roots. But taken as units for measuring

Theory 3: inches and cylinders

An independent set of considerations shows other advantages to the adoption of the English inch in early medieval times and also provides an explanation for the "Assize of London," the strange-looking $31\frac{1}{2}$–63–126–252 sequence of cask measures that has been associated with the wine gallon for centuries and that continues to be embedded in U.S. law to this day. The key lies in Queen Anne's statute of 1706 (Chapter 3) that put into law the traditional definition of the wine gallon in terms of cubic inches. Here's the wording of that statute again:

> [A]ny round Vessel (commonly called a Cylinder) having an even Bottom, and being seven Inches Diameter throughout, and six Inches deep from the Top of the Inside to the Bottom, or any Vessel containing two hundred thirty-one cubical Inches, and no more, shall be deemed and taken to be a lawful Wine Gallon; and it is hereby declared, That two hundred fifty-two Gallons, consisting each of two hundred thirty-one cubical Inches, shall be deemed a Ton of Wine, and that one hundred twenty-six such Gallons shall be deemed a Butt or Pipe of Wine, and that sixty-three such Gallons shall be deemed an Hogshead of Wine.

It's apparent that the authors of the statute of 1706 believed that they were precisely defining a single standard. But if you do the math implied by the statute, then a wine gallon is either "two hundred thirty-one cubical inches, and no more," or it is the volume of a cylinder 6 inches deep and 7 inches in diameter, which is actually $230.90706\ldots$ in^3. We see this discrepancy (a difference of considerably less than one part in a thousand) because nowadays we typically use the extraordinarily precise value of the constant π built into our calculators. But every educated person from the time of Archimedes right up to the invention of the electronic calculator (this includes your author, who remembers it clearly) assumed the value of π to be $\frac{22}{7}$ (i.e., $3\frac{1}{7}$), which is exact enough even today for any but the most critical applications. Indeed, the assumption "that the Proportion between the Diameter and the Circumference [of the bushel measure] was as seven is to twenty-two" was explicitly stated in testimony before the Committee of 1758.[15] And if π is $\frac{22}{7}$, then the two definitions given in the statute are identical.

The existence of an inch standard and an assumed value for π that makes possible the definition of a gallon as a cylinder measured in exact whole numbers of inches also makes it possible to define the cask standards referenced in the statute in exact whole numbers of inches,[16] and the system then looks like the following:

Unit	Gallons	Diameter (in.)	Height (in.)
barrel	$31\frac{1}{2}$	21	21
tierce	42	21	28
hogshead	63	21	42
tertian	84	21	56
pipe[17]	126	21	84
tun	252	42	42

With diameters equal to their heights, the barrel and the tun are to cylinders what cubes are to rectangular containers (see Figure 6.2).

The Parliamentary Commission of 1841 reported that

> much stress has been laid, in the opinions given to us, on the importance of requiring that all wooden measures (casks excepted), and all measures used for sales in the open air, be cylindrical, with their diameter equal to their depth.... The principal reason for urging it appears to be, that, as a single dimension would define the capacity of a measure, it would afford great facility for the hasty examination of measures used in fairs and other public places to which it is inconvenient to carry the local standards for comparison...[18]

The Commissioners rejected this suggestion, favoring instead the use of a special gauge constructed for the purpose, but the many requests for measures with diameters equal to their heights makes it reasonable to think that earlier generations lacking special gauges might have thought highly of the

Chapter 6. King Henry's arm

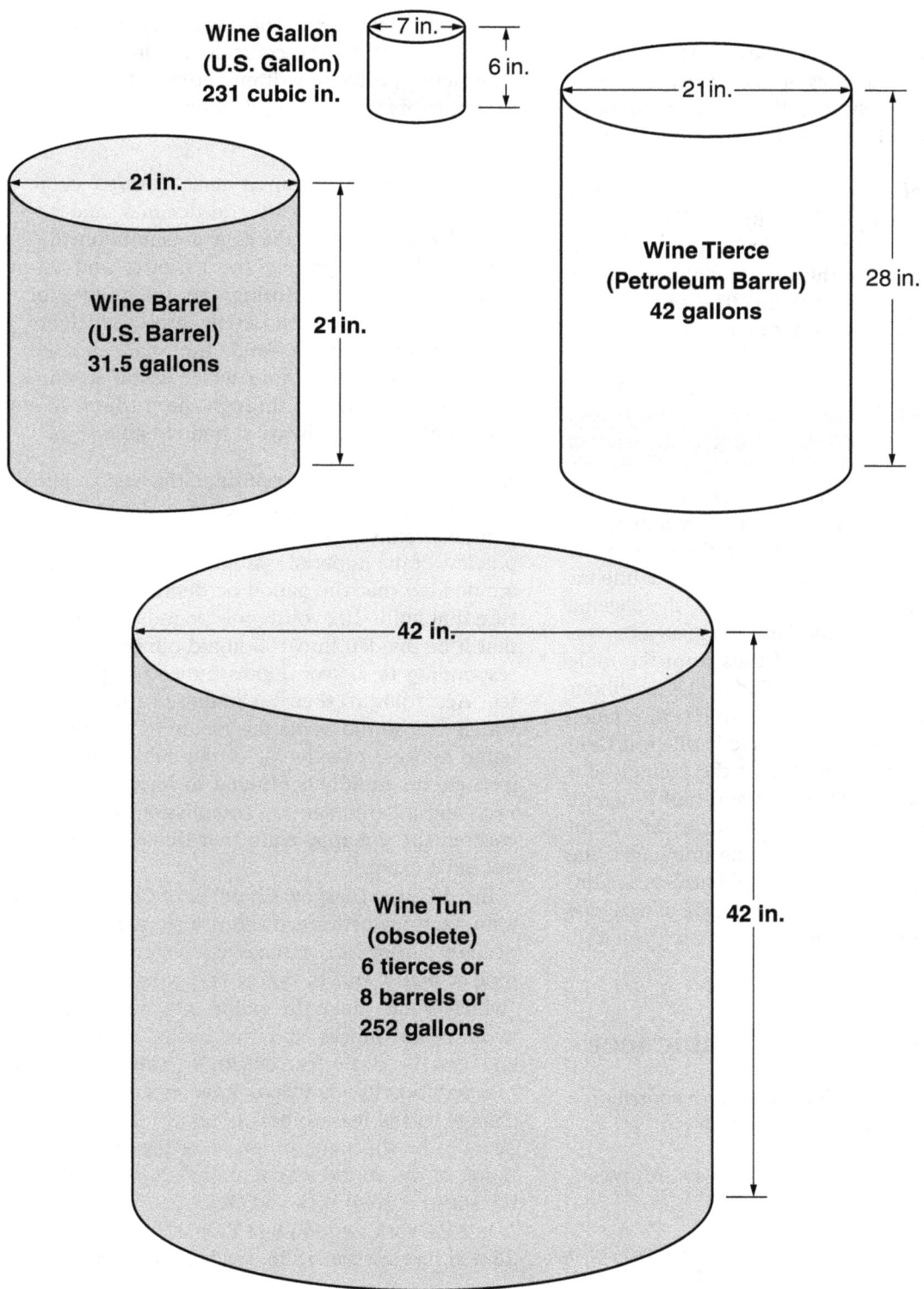

Figure 6.2: Cask system implied by the statute 6 Anne c. 27 §22 (1706). The measurements are exact if $\pi = \frac{22}{7}$ as assumed by the statute

idea.

Cylindrical equivalents provide at least notionally a way to check the accuracy of cask measures,[19] and they also make it easy to estimate the amount of grain stored in a cylindrical granary. The only simple method ever devised for doing this is to measure the depth of grain with a measuring rod and then somehow calculate the volume. The cylindrical equivalents built into the definition of the Queen Anne wine gallon make this easy to do in granaries built to take advantage of them. The same would apply to water and sewer pipes measuring 7 or 14 inches in diameter, though these are not now much in evidence. If a gallon is a cylinder with a diameter of 7 inches and a height of 6 inches, then every foot length of a pipe with an inside diameter of 7 inches would contain 2 gallons and every foot length of a pipe 14 inches in diameter would contain 8 gallons.

It's tempting to theorize that these cylinder relationships had something to do with the decision of Peter the Great of Russia (1682–1725) to define his country's imperial standard of length, the sagene, as 7 English feet. The imperial Russian sagene was 84 of our inches, exactly, and thus from the table above it will be seen that our U.S. barrel is a cylinder whose height and diameter are both (by the Queen Anne definition of the gallon) exactly one fourth of a sagene, and that the old tun is the volume of a cylinder half a sagene in height and half a sagene in diameter. One third of a sagene, or exactly 28 of our inches, was an imperial Russian unit called the archine; so today's international oil barrel is a cylinder one archine in height and one-fourth of a sagene in diameter. This seems somehow fitting given Russia's status as a producer of oil.

Theory 4: ounces and cubic feet

Yet another relationship that could have contributed to the definition of the English foot is

> 1 cubic foot of water weighs 1000 avoirdupois ounces

This relationship was close enough to have been declared exact in an inquiry of 1749, which also noted that 1000 ounces ($62\frac{1}{2}$ avoirdupois pounds) is not a bad estimate of the weight of wheat in a Winchester bushel.[20] Thomas Jefferson was particularly struck by the coincidence; in a four-page manuscript probably written between his tenure as Secretary of State and his election to the presidency, he wrote:

> 1000. ounces avoirdupois make exactly a cubic foot of water. This integral decimal, and cubical relationship induces a presumption that while deciding among the varieties and uncertainties which, during the ruder ages of the arts, we know had crept into the weights and measures of England, they had adopted for their standard those which stood so conveniently connected through the medium of a natural element, always at hand to appeal to.[21]

A commission of 1814 confirmed the basic relationship, finding that a cubic foot of water at $56\frac{1}{2}$ degrees Fahrenheit weighed 1000 ounces exactly. In a preview of the Imperial system, the commission recommended that the gallon be defined as the measure that holds 10 avoirdupois pounds of water and that it be divided into 160 liquid ounces, each corresponding to 1 avoirdupois ounce weight of water. According to their reckoning of the weight of water, this would make the resulting half pint (10 liquid ounces) exactly $\frac{1}{100}$ of the cubic foot.[22] But instruments rapidly continued to improve in accuracy, and a Parliamentary commission of 1819 discovered the unhappy truth that this relationship is not quite exact.[23]

In 1823, the Glasgow Chamber of Commerce and Manufactures petitioned the British government to declare a thousand ounces equivalent to a cubic foot of water and to define the gallon as 270 in^3 (which would make the gallon of water weigh 125 avoirdupois ounces and the bushel of 8 gallons equal to $1\frac{1}{4}$ cubic feet exactly). This appeal was rejected because it would have required a slight change to the avoirdupois pound.[24] But what appears to be the same proposal as regards the definition of the pound was actually adopted by law in the states of New York and Ohio.

In New York, Section 8 of Title II, passed in 1827–28 and reprinted in 1836, reads as follows:

> The unit or standard of weight, from which all other weights shall be derived and ascertained,

shall be the pound, of such magnitude, that the weight of a cubic foot of distilled water, at its maximum density, weighed in a vacuum with brass weights, shall be equal to sixty-two and a half such pounds.[25]

The same New York law of 1827–28 attempted to institute an analogue of the Imperial gallon by defining the gallon as "a vessel of such capacity, as to contain, at the mean pressure of the atmosphere, at the level of the sea, ten pounds of distilled water, at its maximum density."[26] Like the Imperial gallon, also defined as containing 10 pounds of water, the New York gallon was to serve for both liquid and dry substances, thus replacing both the wine gallon and the Winchester gallon, though it differed slightly in capacity from the Imperial gallon due to the fact that it was based on a slightly different avoirdupois pound in order to align exactly with the cubic foot. This statute was almost immediately replaced (April 1829) with another defining separate liquid and dry gallons, the one for liquids to contain 8 pounds of water and the other, 10 pounds of water.[27] The same system was instituted in Ohio by an almost identically worded law adopted in 1835.[28]

This curious law stayed on the books in Ohio until 1846 and in New York until 1851, when the state standards were declared to be the same as those adopted by a joint resolution of the U.S. Congress in 1836 (wine gallon of 231 in^3, Winchester bushel of 2150.42 in^3). It would be interesting to know whether any official standards were actually promulgated in accordance with these definitions; I doubt it.

The British themselves took the other route, keeping the avoirdupois pound as it was and abandoning an exact simple relationship between the gallon and the cubic foot.

While not exact, the close relationship between the cubic foot and a water weight of a thousand ounces does provide an excellent working approximation. Furthermore, a water-weight of $62\frac{1}{2}$ pounds per cubic foot is easy to work with in a basically decimal system of larger weights, because it doubles to give factors of 125, 250, 500, and so on. For example, if the cubic foot contains 1000 avoirdupois ounces of water, then 32 cubic feet of water weigh exactly one ton of 2000 pounds.[29] So the cubic English foot turns out to be very handy in figuring the displacement of ships due to different weights of cargo.

There even appears to be evidence of a more general tradition here. A thousand *troy* ounces of water occupies the cube of another major foot standard, the Rhineland foot, which was the premetric length standard of Prussia and a number of other German states.[30] And a thousand troy ounces of wine (on the standard of the wine gallon) occupies a cubic Greek foot,[31] giving us yet another example of the 64–80–100 system discussed in Chapter 4:

1 cubic Greek foot contains
    640 troy ounces of barley
    800 troy ounces of wheat
    1000 troy ounces of wine

It follows from this that 25 U.S. gallons occupies 3 cubic Greek feet.[32]

The relationship between the cubic English foot and the avoirdupois ounce is so close, so useful, and so historically reasonable that several early writers on the subject (for example, Barlow, Reynardson, and Nicholson) took it as proof that this equivalence was the original basis of both standards.[33] Reynardson and Nicholson argued at length that the English foot was based on the avoirdupois water weight, assuming incorrectly that avoirdupois goes back in England as far as troy. Indeed, according to Reynardson, most authors of his time (1749) believed that avoirdupois weight came directly from the Roman occupation of Britain. But while the avoirdupois standard is unarguably Roman in origin, there is no physical evidence of the use of avoirdupois weight in England prior to Edward I, and, as stated earlier, modern opinion traces the introduction of avoirdupois to medieval trade with Florence.

Direct inheritance of avoirdupois from the Roman occupation never was likely; as Henry justly observed long before the discovery of Saxon graves proved the point,

> There is no probability... in the conjecture of some learned men,—that the Anglo-Saxons adopted the Roman weights and measures which they found in use among the provincial Britons, and laid their own [brought over from

Chapter 6. King Henry's arm

Germany] aside. This was a compliment they were by no means disposed to pay, to a nation with whom they had no friendly intercourse, and against whom they were animated with the most implacable hatred.[34]

What has been found from the Roman occupation is, not surprisingly, the Roman pound. In the 1930s, a perfectly preserved Roman libra with the Greek inscription "Libra 1" was discovered at the site of a Roman villa in North Lincolnshire and found to weigh 5051.25 troy grains.[35] This would, of course, have been divided into 12 ounces, and it's worth noting as yet another instance of the interconnectedness of these ancient standards that 16 of these ounces would give a very good approximation to the tower merchant's pound of 6750 grains.

Since the available evidence puts the origin of the English length measures a couple of centuries earlier than the first widespread use of the avoirdupois weight standard in England, it can't be that the weights gave rise to the lengths. And it's beyond question that the weight standard we call avoirdupois had been in use for over a thousand years before it got to England, so the length standard (found nowhere else and unknown before the 10th century A.D.) certainly didn't give rise to the weights.

While the connection between avoirdupois and the cubic foot may well have cemented the role of both, it's unlikely that it can provide a suitable explanation for the origin of the English length standard by itself.

Assessing the theories

We can sum up the preceding discussion with the observation that the establishment of the English inch, foot, and yard enabled the use of a number of more or less independent shortcuts that could roughly bridge the wine gallon, the Winchester bushel, and three different systems of weight. How or whether this capability actually defined the English length standard remains an open question.

While the absence of further evidence makes it difficult to prove the proposition that the later English length standard was created in pragmatic relation to the existing weight and capacity standards,

this lack of evidence does not, oddly enough, count against it. At the risk of an imputation of the *argumentum ad ignorantiam*, I have to observe that if such relationships were in actual use, an absence of evidence is just what we would expect to see. This is because measurement was considered valuable advanced technology, and the part that applied to any given trade was among the secret techniques or "mysteries" of the guild.

Take, for example, the larger cask measures discussed in Chapter 3. The construction of the casks themselves was the domain of the coopers (makers of barrels). The statute 23 Henry 8 c. 4 §1 (1531) commanded that no brewer of ale or beer "shall frome hensforth occupie... the misterie or crafte of Cowpers" by making their own vessels for selling the ale or beer. So it was entirely up to the coopers to see to it that their wooden vessels actually complied with the laws regarding their contents in gallons and who would have made use of any useful relationships between the capacity and length measures. These rights were maintained till the 1750s.[36]

Similarly for the yard itself. According to Connor,[37] "records exist... showing that the custody and use of yard measure for the purposes of trade was jealously guarded by the cloth guilds." A royal charter of 1439 gave the Merchant Taylors' Company the exclusive right "when the said Maister shall think beneficial to the said Mysterie" to check standards "throughout the whole cittie [of London] and the suburb thereof" using their silver yard (mentioned above) and other standards maintained by them alone.

Weights, too, were under the control of the guilds. The Goldsmiths' Company was chartered to supervise London's tower and troy weights from 1392 to 1679,[38] and the statute 2 Henry 6 c. 15 (1423) required that the Assayer and Controller of the Mint itself be "credible, substantial, and expert men, having perfect knowledge in the mystery of goldsmiths and the Mint" (*crediblez & expertz persones aiantz notoier science en le mistier & dorfeo^r & de Mynt*). The sizing and marking of "all manner of Brass Weights, to be made or wrought, or to be uttered, or kept for Sale, within the City of London, or three miles from the same" was by charter of James I (September 1614) the province of the

Founders Company of London,[39] a monopoly they maintained until 1889.[40] Another charter of James I (April 1611) gave the Plumber's Company authority "to search, correct, reform, amend, assay, and try, amongst other Things, all Weights of Lead within the City of *London,* Suburbs thereof, and within seven Miles of the same City, with Powers to enter into Houses, Shops, &c. and to search, and seize such Weights as should be found insufficient or deceitful."[41]

Given the closeness with which the guilds held both the standards themselves and the right to enforce compliance to them, together with the fact that most knowledge in those days was conveyed orally, it's not surprising that we don't know more about their methods. Until some new information comes forward, however, the explanations given here for the origin of the English length measures must remain conjectural.

Notes

[1] In early manuscripts the word *ell* meant the unit we now call the yard. The cubit of 18 inches appears in most dictionaries of English, but in fact the cubit has never been used in English-speaking countries as a measure of length; it exists in English only to render the unit in translations from other languages. See the Glossary beginning on page 139 for more about the names of these units.

[2] CONNOR, 82–84. The examples of the modern English foot found in Petrie's study of English building measurements don't date back farther than the 12th century (*Inductive Metrology,* 106).

[3] As the yard itself was used primarily for cloth measure, neither of the two earliest examples is marked in inches, but rather in binary divisions of the yard. The 1445 yard is divided into four nine-inch sections (that is, four spans), the last of which is divided into a half span, or eighth of a yard, and two nails, or sixteenths of a yard, while the 1497 yard is divided into $\frac{1}{2}$, $\frac{1}{4}$, $\frac{1}{8}$, $\frac{1}{16}$, and $\frac{1}{32}$ of a yard (CONNOR, 234–35). The most important binary division of the yard, the nail (sixteenth of a yard or $2\frac{1}{4}$ inches), can still be found marked on yard measures embedded in the counters of American fabric stores. Notwithstanding the absence of foot and inch markings from these early yards, statutes dating back to 1305 (SKINNER, 94–95; CONNOR, 322) show that the yard was from the first considered to be made up of 3 feet of 12 inches each. Following a very old tradition, each inch was conceived of as the length of 3 barleycorns laid end to end, which is not unreasonable for real grains of barley but was also applied to other traditional inch standards. The oldest Exchequer yard (Henry VII, 1497) is marked for $\frac{1}{16}$, $\frac{1}{8}$, $\frac{1}{4}$, $\frac{1}{3}$, and $\frac{1}{2}$ yard from one end and 1 foot divided into 12 inches on the other (CONNOR, 239); see Figure 6.1 (from Chisolm, *Weighing and Measuring,* 51).

[4] *Mercatorum falsam ulnam castigavit; brachii sua mensura adhibita, omnibusque per Anglium proposita.* CONNOR, 83 (Connor's translation).

[5] *Minutes of Evidence* (1862), 2.

[6] SKINNER, 95.

[7] CONNOR, 82.

[8] Huggins, "Anglo-Saxon Timber Building Measurements," 24.

[9] The theory is due to Huggins ("Anglo-Saxon Timber Building Measurements," 26), who advances it only as a possibility.

[10] This line of thought was inspired by an observation made by John Quincy Adams in his report to Congress, though he did not carry it through to the conclusion suggested here.

[11] See Appendix A for more on this.

[12] The connection between tower weight and the cubic English foot appears to have been first observed in modern times by Adams (27, 33, 35).

[13] Indeed, Skene Keith (*Observations on the Final Report of the Commisioners of Weights and Measures,* 44) suggested in 1822 that the bushel simply be redefined as 2160 in^3.

[14] Nichols, *Business Guide,* 380.

[15] *Report from the Committee* (1758), 45.

[16] I believe that I am the first modern commentator to notice this.

[17] While not significant in itself, it's worth noting in this context that the dimensions of the cask unit called the *pipe* under this interpretation would provide a simple explanation of its name.

[18] *Report of the Commissioners, 1841,* 14.

[19] "Notionally" because casks were never actually shaped like cylinders (they are wider in the middle than at the ends). Like all medieval guilds, that of the coopers had its own secrets, and the relationship (if any) of the system described here to their practice is a subject for further investigation.

[20] Reynardson, "A State of the English Weights and Measures of Capacity", 63. A weight of $62\frac{1}{2}$ avoirdupois pounds of wheat per Winchester bushel is actually a bit on the high side but not impossible; see Appendix A.

[21] Boyd, *Papers of Thomas Jefferson,* 16:613.

[22] *Report of the Select Committee Appointed to Enquire into the Original Standards,* 5; CONNOR, 252. Writing in 1749, Reynardson ("A State of the English Weights and Measures of Capacity," 64) noted that this is the same as saying that a cube $\frac{1}{10}$ of a foot on a side holds water weighing 1 avoirdupois ounce, and Thomas Jefferson proposed in 1790 that the avoirdupois ounce in the U.S. actually be defined to be "of the weight of a cube of rainwater, of one tenth of a foot" (Boyd, *Papers of Thomas Jefferson,* 662).

[23] *First Report of the Commissioners Appointed to Consider the Subject of Weights and Measures,* 4. The commission recommended that reform begin "by declaring, that nineteen cubic inches of distilled water, at the temperature of 50°, must weigh exactly ten ounces Troy, or 4,800 grains," which would make the cubic foot hold about 997.8 avoirdupois ounces. The Sale of Gas Act (1859) declared a cubic foot of water at 62 °F to weigh 62.321 pounds avoirdupois or 997.136 ounces (*Memorandum on the Standard Bushel Measure,* 2).

[24] *Report of the Select Committee of the House of Lords,* 9.

[25] *Revised Statutes of the State of New-York* (1829), 1:607; *Revised Statutes of the State of New-York* (1836), 1:617.

[26] *Revised Statutes of the State of New-York* (1829), 1:608.

[27] *Revised Statutes of the State of New-York* (1829), 3:315.

NOTES

[28] Swan, *Statutes of the State of Ohio,* 986–87.

[29] Barlow, W., "An Account of the Analogy betwixt English Weights and Measures of Capacity," 457; also ADAMS, 126.

[30] According to the definition of the Imperial bushel used for most of the 20th century (see note on page 65), a cubic inch of water at 62 degrees Fahrenheit weighs 252.28589 grains. A thousand troy ounces (480,000 grains) would therefore occupy a cube with sides 12.39128 inches in length. The 19th century Rhineland foot in Berlin was 31.38535 cm (12.35643 inches) and in Amsterdam was 31.39465 cm (12.3601 inches). A colonial variant of the Dutch version of the Rhineland foot survived into the mid-20th century as the Cape foot of South Africa at 31.486 cm (12.396 inches). The Scotch foot, which appears to come from the same tradition, is put by various sources at 12.353–12.400 inches. Nicholson discusses the relationship between the Rhineland foot and a thousand troy ounces at length. It's my opinion that the Rhineland foot is 12 "inches" of the 20 units into which the Egyptian royal cubit (20.62–20.63 inches) was divided for practical purposes; this would put the foot at 12.372–12.378 inches.

[31] If a wine gallon (231 in^3) of wine weighs 10 troy pounds (see Chapter 3), then 1000 troy ounces of that wine occupies 1925 in^3, which is a cube with sides 12.4397 inches in length. The ancient Greek foot was 12.44 inches.

[32] The relationship may not be just theoretical. A survey of official U.S. measures conducted under the direction of John Quincy Adams found (ADAMS, 140-a) that the half-bushel standard of the U.S. Customs House at Bath, Maine, in 1819 had a capacity of 962.5 in^3; the corresponding bushel was therefore 1925 in^3, which is a cubic Greek foot (see preceding note). The capacity of the Customs House standard was calculated from measurements that found its diameter to be 14 inches (top and bottom) and its depth to be 6.25 inches (reported as a decimal, as with all the Customs House standards in the survey). Given the well-known dimensions of the wine gallon (7 inches in diameter and 6 deep), it is likely that the diameter was intended to be 14 inches exactly, and 6.25 is too different from 6 to be a mistake.

[33] Barlow, "An Account of the Analogy betwixt English Weights and Measures of Capacity," 457–58; Reynardson, "A State of the English Weights and Measures of Capacity," 54ff.; Nicholson, *Men and Measures,* 48–57.

[34] Henry, *History of Britain,* 250–51.

[35] *Antiquaries Journal* 13:57–58.

[36] Zupko, *British Weights & Measures,* 85.

[37] CONNOR, 234–35.

[38] Zupko, *British Weights & Measures,* 83.

[39] "An Account of a Comparison lately made by some Gentlemen of the Royal Society," 554. See also *Report from the Committee* (1748), 39.

[40] Zupko, *British Weights & Measures,* 85.

[41] *Report from the Committee* (1758), 37.

Chapter 7

Conclusion

The enormous age of our Customary standards of weight, capacity, and length will, I hope, have inspired some respect for these elements of our cultural tradition, and such considerations would lead naturally to a defense of the future of the Customary System in an "age of metrication." But I will disappoint the reader who is passionately opposed to the metric system by saying that I don't think there's anything left to get excited about.

Metrication: a mixed bag

The last several decades of debate and attempts to get people to change their ways demonstrates that campaigns to effect reform in the area of weights and measures have about the same outcome as campaigns to reform language: the change is just enough to make things more complicated, because people hold on to what's really useful (or hard to change) while adopting just the new bits that make the most sense.

Take our measures of capacity, for example.

The traditional system of capacity measures persists in the U.S. for cooking because it is more user-friendly than its metric equivalent. This is demonstrated by the adaptations that the metric system has had to make in the kitchen in order to remain usable. Many cooks working in the metric system use copies of the Customary teaspoon (at 5 mL) and tablespoon (15 mL) plus clumsy, rough approximations to the traditional measures at a centiliter (100 mL), $\frac{1}{8}$ liter (125 mL), $\frac{1}{4}$ liter (250 mL), $\frac{3}{8}$ liter (375 mL), and $\frac{1}{2}$ liter (500 mL). Some metricated cooks have it even worse; a set of Canadian measuring cups from China is marked 1 cup (25 cL), $\frac{1}{2}$ cup (12.5 cL), $\frac{1}{4}$ cup (6 cL), and $\frac{1}{8}$ cup (3 cL). Compare the series 25, 12.5, 6, and 3 cL with 8, 4, 2, and 1 fluid ounces. This is progress?

It is said that European cooks weigh many things that we in the U.S. measure out by capacity because they are more concerned with accuracy, but given the much greater ease and efficiency of measuring by capacity, it is more likely that the Europeans have resorted to measuring by weight because they have lost the natural, easy, close-spaced sequence of measures that typified virtually all the premetric systems of measures and allowed almost all recipes to be specified in simple round numbers of units.

Capacity measures are a nice demonstration of the truth that factors of 10 are useful for some things and not so useful for others. For lengths, 10s make a lot of sense, but for capacities, a factor of 10 is too big a stretch; we want unit measures spaced more closely, and most cultures prior to the implementation of the metric system settled on divisions into halves and thirds.[1] Metric users must try to recreate the utility of these ideal intervals by dividing the decimal base in ways that are ungraceful, to say the least.

Look at the mess that metrication has made in containers for wine and spirits. Here we have a nice, neat, "scientific" unit for liquids, the liter, with a nice, neat weight equivalence (a liter weighs a kilogram, more or less). So do we sell wine and spirits by the liter and half liter? Why no, we sell it by $\frac{3}{8}$, $\frac{3}{4}$, $1\frac{1}{2}$, and even, God help us, $1\frac{3}{4}$ liter (or 375, 750, 1500, and 1750 mL). And when we talk about the weight of the wine in a case of wine (say), we get to work with relationships like 12×750 grams = 9 kilograms (assuming that wine weighs the same as water), which isn't a nice round number in anyone's system.[2] This is clearly a step backward from the system we used to have in this country, when wine and spirits were sold by the fifth of a gallon and a case of fifths contained 10 avoirdupois pounds of liquid.[3]

Some Americans seem to be under the impression

that the 750 mL wine bottle is a local adaptation that wouldn't be seen in countries properly using the metric system; surely, they say, those more advanced peoples must sell their booze by the liter. But I haven't seen a country in the world where wine and spirits are predominantly sold by the liter, because *it's the wrong size* for wine and spirits. Unless you're an unusually large person, the liter doesn't comfortably fit the hand, which becomes an increasingly important consideration the more you've had to drink. That's why we've settled on 750 mL for the bottle itself and 1.5 or 1.75 liters for jugs with handles, which is ergonomically the right thing to do but no improvement on the Customary measures formerly used. It's not an accident that the old wine fifth hung on in England under the name "reputed quart" for a century and a half after it was officially replaced by Imperial measure; indeed, the Commission of 1841 recommended that "the wine-bottle" (as they called it) be made an Exchequer standard.[4] This never happened, but the unit held out anyway simply because it was the right size for a bottle of wine, a size that we now continue as best we can in the form of an unwieldy metric equivalent. The British continued to cling to their "reputed quart" until it was finally prohibited by law in 1963.[5]

Or consider the 12-ounce (355 mL) beer bottle. As you saw in Chapter 4, this particular unit can be traced with fair certainty clear back to the Phoenecians. Why do we continue to use it? Because it's *the right size for a bottle of beer*. For proof, visit any country that uses the metric system, order a bottle of beer, and see what size it comes in. You will almost always get a 33 cL (330 mL) bottle—not, please note, $333\frac{1}{3}$ mL, which is what you'd get if consistency were a real consideration. This is just a lame metric implementation of $\frac{3}{4}$ pint, or the unit of roughly that size in many other traditional systems. You will never get a liter bottle of beer in a restaurant. Sometimes you can order a half liter bottle, but this is not nearly as common as the regular beer bottle (even though the half liter is a very common size for beer sold by the glass). Just as the wine bottle (the fifth) is the right size for a bottle of wine, a whole mysterious set of ergonomics makes 12 fluid ounces the optimum size for a bottle of beer or a can of soda. The Customary system definitely got this right, too.

From the standpoint of both consistency and basic usefulness, therefore, our traditional system of capacity measures is decidedly superior to its metric counterpart, and for that reason it will probably persist into the indefinite future.

Our system of shorter lengths, on the other hand, has little to recommend it. The ancient land measures are fine—poetically simple and just as usable as metric, which is a good thing, because acres are so deeply embedded in U.S. land titles that they will never be replaced by ares and hectares. But our inch, foot, and yard standards are medieval inventions whose reasons for being no longer apply.

The fact is that the shorter metric units of length got it right in a way that the metric units of capacity did not. As you saw in Chapter 1, the meter is virtually identical to a foundational unit of our land measure, the Northern yard, which is a standard far older than the synthetic English yard that metric opponents are ready to man the barricades to defend. And the centimeter is certainly no worse suited to the measurement of smallish objects than the inch. In the absence of a duodecimal (base-12) system of notation, the division of the foot into 12 inches is not a net advantage, and in lengths shorter than an inch, the decimal centimeter and millimeter are far more practical than fractions of an inch expressed in sixteenths or thirty-seconds. There is really no question about this.

Living with the status quo

Unsatisfactory as they are in practice, however, we seem to be past the point where completely abandoning feet and inches will ever be economically worth the trouble. We could phase out fractions of an inch for fasteners (nuts, bolts, etc.) and the wrenches that go with them, and eventually maybe we will, but every American who needs wrenches keeps a set of both SAE and metric sizes already. And even if the sales of SAE socket sizes eventually go to zero, the SAE drives (most commonly, $\frac{1}{4}$, $\frac{3}{8}$, $\frac{1}{2}$, and $\frac{3}{4}$ inch) on which the metric sockets fit in the U.S. will never disappear, because it's simply not worth the trouble to replace them. The same is true for the dimensions of lumber, sheetrock, doors, and other building materials.

But this is not to say that we can't make things easier on ourselves. Keeping inches for the dimensions of building materials, for example, doesn't mean that we have to keep measuring in binary fractions of an inch. This maddening way of working apparently came into use in the U.S. in late colonial times. The Parliamentary yard of 1758 was divided into three feet, one of which was divided into 12 inches, and one of the inches into "ten equal parts,"[6] and Thomas Jefferson, in his 1790 report to Congress on weights and measures, said that the inch was commonly divided into 10 "lines"[7]; whereas the author of an 1823 proposal to Congress called for the yard measure "to be divided with all possible care and accuracy into feet, inches, quarter inches, eighths of an inch, sixteenths of an inch, and thirty-second parts of an inch; such being the division commonly in use among workmen in the United States."[8] Enabling a return to the earlier practice simply by providing tenths of an inch along one edge of all our rulers and tapes would make many ordinary shop calculations vastly easier at virtually no cost.

Similarly with weights. Like the wine bottle in comparison with the liter, the pound is an inherently more appropriate weight for ordinary purposes than the kilogram, which is why you will so frequently see meat, cheese, and other goods priced in metric countries by the half kilo. Over the last decade, while most people weren't really noticing, U.S. groceries have universally gone over to a system of pounds and hundredths or thousandths of a pound. Avoirdupois ounces (as opposed to fluid ounces) are used less and less frequently, and it's not hard to imagine them going the way of the avoirdupois dram. Historically, this would be no loss; as recounted in Chapter 5, ounces came late in the history of avoirdupois weight, which was originally used for quantities no smaller than 7 pounds (the clove or half stone). Life would be both simpler and more in keeping with the underlying basis of the Customary System if all ounces were troy ounces (used only for precious metals and drugs) just as all pounds nowadays are avoirdupois pounds. All we'd have to remember is that a ton is 2000 pounds and that gold and silver are traded by the (troy) ounce, a unit the pound has nothing to do with—which, as you've seen, happens to be historically true.

But unlike many people who have delved deeply into the subject, I feel no particular need to push hard for reforms; we've reached an accommodation that's probably gone as far as it needs to.

Most people don't know this, but the U.S. was the first country in the world to adopt a decimal monetary system, and it could have been the first to institute a complete decimalization of its weights and measures. In the 1790 report mentioned above, Thomas Jefferson ended four years of intensive study undertaken at the request of Congress with a proposal for a completely new system of decimal units the same year in which the French proposed the metric system, and long before they were in a position to implement it. It would have been relatively easy to adopt the new system that year, but despite the support of George Washington, the revolutionary moment passed, and the political climate was never that favorable for such a change again.[9]

So we had our chance for a complete overhaul and we decided to pass. And really, aside from the conceptual pain of a mixed set of units, the current situation isn't that bad—no worse, certainly, than living with a language that arbitrarily mixes French and German. Gallons of milk consort with 2 liter bottles of soda without apparent damage to the social fabric. It would be nice if we could return to using the more convenient Customary sequence for all liquids (it would be worth it just to bring back the gill), but this isn't killing us.

The complaint that a mixture of Customary and metric systems impairs international trade is mostly hogwash. Even the tub-thumping 1971 U.S. Department of Commerce tract *A Metric America: A Decision Whose Time Has Come* had to admit that according to exporters,

> differences in measurement systems and standards seemed relatively unimportant; they put more emphasis on reliability, reputation, price, superior technology, and high quality of product.[10]

The automation of manufacturing processes since then has largely made systems of measurement a non-issue; dimensions of many manufactured goods can be changed just by changing the programming of the machines that create them. Where the measurement system makes any difference, manufacturers simply make one run for sale in metric countries and another for sale in the U.S.

When it comes to food packaging, the difference in measurement systems is often irrelevant. A bag of corn chips obtained for my lunch today claims to contain "$3\frac{3}{8}$ ounces (95.6 grams)." A bottle of tomato juice in my refrigerator is labeled "46 fl. oz. (1.36 L)." These sizes were determined for marketing reasons, and if the Customary designations were eliminated tomorrow, they would still be 95.6 grams and 1.36 liters.

In many cases, ostensible adaptation to a particular measurement system is simply an excuse for manipulating the packaging. This is certainly true for beverages; many of us are old enough to remember when wine bottles shrank from $\frac{1}{5}$ gallon to 750 mL with no corresponding reduction in price. And witness the recent silent switch, now almost complete, from pint juice bottles to ones holding 450 mL — a shrinkage that hands the producer a tidy 5.2 percent increase in profit without fitting the new bottle to a metric distribution system any better than the old one did.

Our system of standards is not where our problem lies; with the help of a relatively painless transition to decimal inches, we can live with what we've got. The big problem is the demise of unit pricing.

Keeping in sight the historical objective

Search far enough into the past and you will find that the reason human beings invented standards of measurement was to keep from being cheated. But here we are, able to measure the world to seven places of precision yet less able to compare prices than we were 200 years ago.

Many states—enough to make this a general practice throughout the U.S.—have laws requiring milk to be sold by the gallon, half gallon, quart, pint, and half pint. The reason they restrict retail sale to these particular packages is to make it immediately clear whether the milk in one container is selling for more than the milk in another container. Similar laws used to require bread to be sold only in certain sizes. The same reasoning applied to all the old statutes we've reviewed in this book that specify the size of barrels for various goods and allowed people to immediately compare the price of one barrel of beef or flour or herring with another.

With the rise of modern packaging, however, this most basic and ancient advantage of standardization has been lost almost entirely. A recent trip down the shampoo aisle of my local supermarket found bottles marked and priced as follows:

Fluid ounces	mL	Price
8.45	250	5.49
10.1	300	3.29
12	355	2.49
12.6	375	3.85
13	384	3.59
13.5	400	2.99
14.2	420	4.79
15	443	0.99
15.9	470	3.85
23.7	700	4.99
23.7	671 gm (!)	4.99
23.7	700	6.99
25.4	750	5.89

This ridiculous hodgepodge of capacities is deliberately intended to make easy price comparisons among brands virtually impossible, and the practice in many stores of providing calculated unit prices in type sizes too small to see confirms the problem without offering a workable solution.

If reformers want to really do us some good, they should advocate a return to standard unit packaging. This would mean, in the case of liquids, requiring all containers to be sized in round-number units of the Customary system—100 gallons, 50 gallons, 20 gallons, 10 gallons, 5 gallons, 2 gallons, gallon, half gallon, quart, pint, cup, gill, fluid ounce, and tablespoon (half ounce)—and optionally allowing them to be sold by major units of the metric system (100 liters, 50 liters, 10 liters, 5 liters, 2 liters, 1 liter, half liter, quarter liter). Plenty of choices there for the package designer, and no bar to international trade.

We could require all packaged dry substances measured by weight to be provided in packages of 100, 50, 25, 20, 10, 5, 2, 1, half, quarter, or eighth of a pound and marked that way (no ounces, though in many cases these packages will be exactly the same size as packages now marked in ounces), optionally allowing the kilo, half kilo, and quarter kilo. All packaged dry substances measured by volume would continue to be sold by Winchester measure (bushel, peck, dry gallon, dry quart, dry pint). This is already the practice in most states for produce sold by volume. Such a policy would restore the ability of consumers to compare prices without seriously affecting any manufacturer that wasn't out to cheat us.

A few diffident recommendations

After more than a quarter century spent studying the subject, therefore, I have just three recommendations:

- Legislate unit pricing for all prepackaged goods as outlined above.

- Require rulers and tapes graduated in inches to be graduated along one edge with inches divided into tenths in all orders for such tools in military, government, educational, and corporate procurement. Begin to abandon binary divisions of the inch in shop classes and make the transition as soon as convenient in individual construction crews. The savings in time here will more than make up for the relatively small transition effort.

- Teach people how their Customary System actually works and stop pretending that it's been completely replaced by the metric system. It's not going to be.

To those who hold religiously to our Customary Units, I say: I understand, but God did not actually give us the units we use now. They weren't provided by Nature, either.

And to those good people who thirst after righteousness in the form of complete metrication, I say: let it go. We've got more important things to worry about.

Notes

[1] Even the Chinese tradition, which was strongly decimal, did not begin to lose its 3s and extra 2s until about the 11th century.

[2] I am aware of the distinction between weight and mass. It's not relevant to what I'm saying here.

[3] Assuming 6 bottles to the case, and remembering that a U.S. gallon contains $8\frac{1}{3}$ pounds of water, exactly.

[4] *Report of the Commisioners, 1841*, 10, 12, 15.

[5] CONNOR, 187 and 297.

[6] *Report from the Committee* (1758), 52.

[7] Boyd, *Papers of Thomas Jefferson*, 632.

[8] Cooper, *Outlines of a Proposed Act of Congress to Regulate Weights and Measures*, 40.

[9] Boyd, *Papers of Thomas Jefferson*, 602–74.

[10] *A Metric America*, 64.

Appendix A. Grain weight as a natural standard

Certain critical features of the history I've recounted in this book have long been obscured by confusion over how to regard the basic definitions of weight and volume given in the old statutes relating to this subject. There appear to be two such definitions. The first definition is contained in two statutes titled *Assisa Panis et Cervisie* (Assize of Bread and Ale), traditionally assigned to 51 Henry 3 (1266), and *Tractatus de Ponderibus et Mensuris* (Treatise on Weights and Measures), traditionally referenced as 31 Edward 1 (1301). As set out in Chapter 4, my reading, and that of several other researchers, is that these two statutes legally establish the 15-ounce tower merchant's pound of 6750 troy grains and define the Winchester bushel based on it as the measure that holds 64 such pounds of wheat. The second basic document is the statute 12 Henry 7 c. 5 (1496), which establishes the troy pound and defines the wine gallon based on it as the measure that holds 8 troy pounds of wheat (Chapter 3).

This much appears beyond dispute. But the same statute of 1496 also defines a bushel of 8 wine gallons, the actual use of which for dry substances remains an open question.

The two sets of definitions are highly consistent; if a given sample of wheat weighs 8 troy pounds per U.S. gallon, then 64 tower merchant's pounds of that same wheat would make a bushel measuring 2165.625 in^3, which is just 0.7 percent larger than our actual bushel. Or, to put it the other way, if a U.S. bushel of wheat weighs 64 tower merchant's pounds, then 8 troy pounds of that wheat will occupy a gallon of 229.38 in^3, which is 0.7 percent smaller than our actual gallon.

Some historians, however, have argued that the pound referred to in the *Tractatus* is actually the troy pound of 5760 troy grains, and others have argued that the bushel referred to in the statute of 1496 is actually the Winchester bushel; in other words, in either case, that the Winchester bushel is the measure of 64 *troy* pounds of wheat. This theory is certainly understandable because that's what 12 Henry 7 c. 5 (1496) seems clearly to say, and, as you will see, it was carried forward for centuries by authorities on the subject; but as far as we can know based on the evidence (or lack of it), the bushel of 64 troy pounds can only be characterized as a fiction. In addition to the fact that there seem to be no recorded instances of such a measure actually existing, it turns out that the definition is physically impossible for real wheat of the quality that we must assume here.

Establishing this basic fact about wheat is not only necessary to some critical determinations regarding the history of our Customary System; it is also the occasion for addressing two important questions, the answers to which illuminate the entire inquiry:

Is the average bulk weight of wheat over time stable enough to actually serve as the basis for a system of weight and capacity?

Which bulk weight of wheat should be used in estimating the capacities of traditional measures based on their legal definitions?

And to explain the pattern of ratios noted in the text, it would also be helpful to get a definitive answer to the question:

What is the simplest ratio between the bulk weights of wheat and barley?

We know a few things for sure going into this. One is that the people who wrote the two sets of English statute definitions clearly *thought* that they were defining a standard of capacity based on a particular weight of wheat. And in fact traditional systems of measurement the world over very commonly do define capacity standards based on the weight of the substances they contain. In some places, such as India, there really were no traditional capacity measures per se, the names of capacity units simply

Appendix A. Grain weight as a natural standard

being a shorthand for certain weights (as, for example, the definition of 60 pounds as a bushel of wheat for legal purposes in the U.S.).

Another thing we know is that the bulk weight of grain, which agronomists and grain inspectors call its *test weight*, has always been the most important of the criteria (leaving aside freedom from contamination by mice and insects) that go into grading it for quality. The reason is simple: test weight is an excellent indicator of the soundness and basic health of grain. And in the case of wheat, it's a good indicator of the quality of the bread that can be made from it. As an old agronomy textbook puts it, "The usual and commercial standard of quality in wheat is the weight per bushel, high weight being associated with qualities desired by the miller."[1]

Seed grain behaves differently depending on test weight, too. Tests carried out at Kansas State University in 1950 demonstrated that samples of high test weight wheat had 1.4 to 2.1 times the germination rate of seeds from low test weight wheat, showed 20 to 40 percent better field emergence, emerged 4 to 6 days sooner, and yielded 5 more bushels an acre.[2]

Because it is the primary standard of quality, test weight has always been a key factor in determining the *price* of a given lot of grain. A New York state law of 1788 required that

> the standard weight of wheat brought to the city of New York for sale, shall be sixty pounds nett to the bushel; and in all cases of sales of wheat in the said city by the bushel, if the same shall exceed the standard weight, the buyer shall pay a proportionably greater price; and if the same shall be less than the said standard the buyer shall pay a proportionably less price...[3]

An almost identically worded Connecticut law of 1796 applied the same requirement to that whole state.[4]

In evidence taken before a British Parliamentary Committee in 1862, expert testimony was given that "quality [of wheat]... can only be determined by reference to the measure in connexion with the weight."[5]

Grading, therefore, has a lot to tell us about the weight of grain.

Here are the minimum (not average) weight requirements for winter wheat, durum wheat, and most varieties of spring wheat graded by the U.S. Department of Agriculture:[6]

Grade	Must weigh no less than
U.S. No. 1	60.0 lb/bu
U.S. No. 2	58.0 lb/bu
U.S. No. 3	56.0 lb/bu
U.S. No. 4	54.0 lb/bu
U.S. No. 5	51.0 lb/bu

The grades below No. 2 are not generally used in products for human consumption. The average weight of wheat *within* each grade will, of course, be above these minimums. It's a melancholy fact about wheat that much more is downgraded by the presence of mouse droppings, insects, and dead birds (there are specified limits for each of these) than by low test weight. Data I obtained in 1985 for three years of U.S. total wheat exports (see page 119) showed the following averages for all export wheat types (winter, spring, white, durum, etc.):

Grade	Mean total export weight
U.S. No. 1	61.03 lb/bu
U.S. No. 2	61.06 lb/bu
U.S. No. 3	60.57 lb/bu
Test grade	59.40 lb/bu

Generic wheat and barley

Before we get any further into the question of grades, it will be well to take a look at the average behavior of wheat and barley weights over a suitably large and varied area to establish a baseline. And it would also be a good idea to step around any questions about the extent to which modern breeding and genetic manipulation has changed basic qualities of grain by looking for data from a time before such practices became widespread.

The best data I can find that meets all of these requirements is a set of U.S. government statistics published in annual reports covering the period 1902–1930.[7] These are for the entire country's wheat production and, starting in 1910, the entire country's barley production—an enormous tonnage—all grades, all regions (Figure A.1).

Appendix A. Grain weight as a natural standard

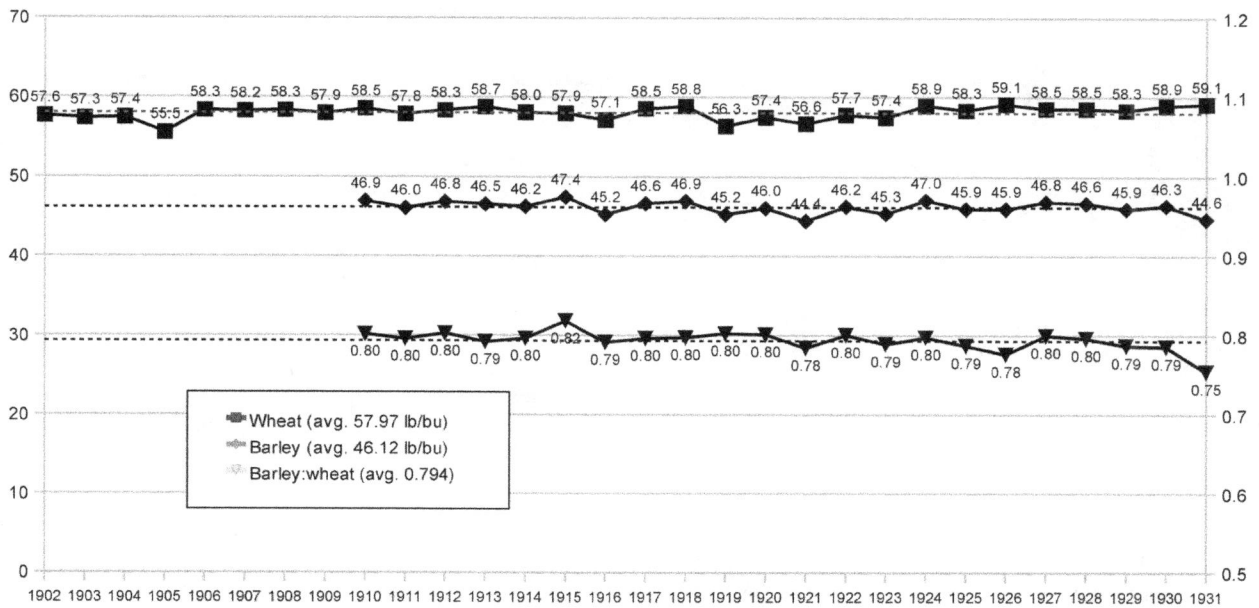

Figure A.1: Test weights of U.S. wheat (1902–1931) and barley (1910–1931), including 10–15% not marketable

The top line shows the average test weight of U.S. wheat for the 30-year period 1902–1931. (The abbreviation lb/bu used throughout this discussion means avoirdupois pounds of 7000 troy grains per Winchester bushel of 2150.42 in^3.) This average wheat averaged 57.97 lb/bu over 30 years of national production data. Average barley, on the second line, averaged 46.12 lb/bu during the 21 years for which data is available. These overall averages are indicated by the dotted lines in the graph. It can be seen that the annual averages for average grain sampled over a large area rarely wander far from the mean, even though there were sizable fluctuations in production due to weather conditions each year. Note, however, that the average weights given here are for *all* wheat and barley, including 10 to 15 percent that was of such low quality that it couldn't even be sold for animal feed, so while these data provide a good demonstration of the stability of test weight on a national basis, they do not answer the question of what weight would have been considered acceptable for wheat of the best quality. More on that shortly.

The lowest of the three lines, plotted against the axis on the right, is the ratio of average weight of wheat to average weight of barley each year for which both sets of weights are available. The average of this ratio for all U.S. wheat and barley over the two decades of production is 0.794; this differs from the $\frac{4}{5}$ barley:wheat ratio discussed in the text by about 0.8 percent.

Another interesting set of data for what we might call "heritage grain" comes from annual plantings at the government Experimental Farm in Ottawa from 1891 to 1911.[8] As in all data from experimental stations, the figures are based on samples weighing no more than a few pounds and therefore display statistical variability at the opposite end of the spectrum from the national averages given in the previous set.

As shown in Figure A.2, there are actually two sorts of wheat and two sorts of barley to keep track of if we look into this closely. Winter wheat is planted in the fall, lies dormant all winter, and then grows in the spring and summer, whereas spring wheat is planted in the spring. In Ottawa, at least, this gives winter wheat a head start, and consequently it's a bit heavier on average (61.3 lb/bu here vs. 59.3 lb/bu for spring wheat). However, in marginal locations like Ottawa, where rains can

Appendix A. Grain weight as a natural standard

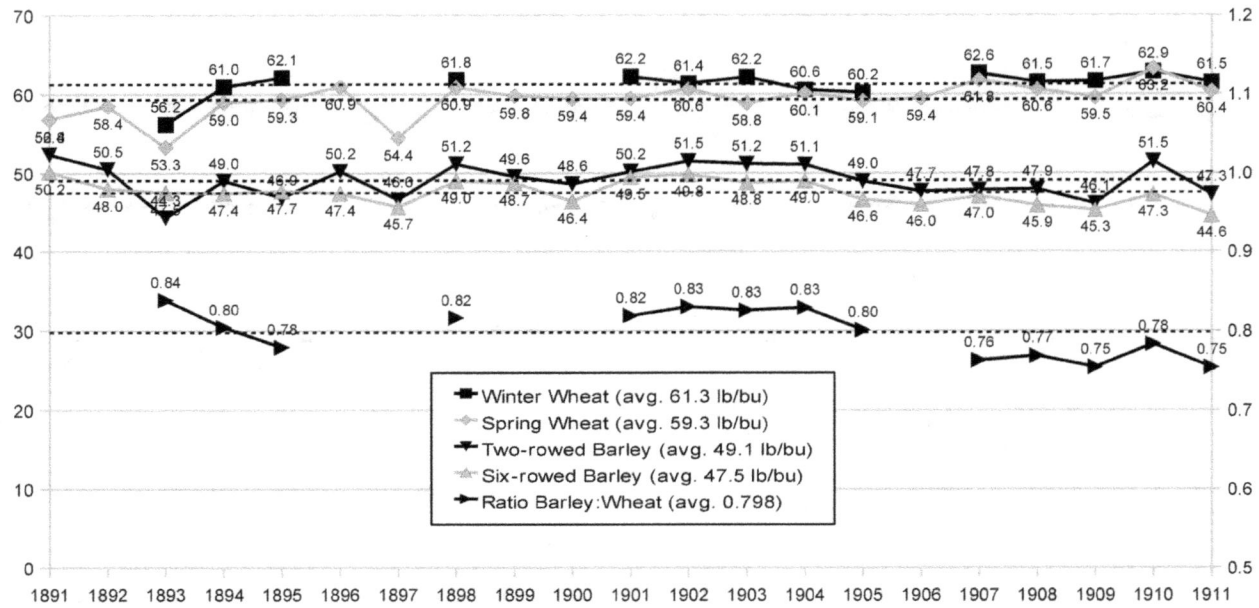

Figure A.2: Wheat and barley at the Experimental Farm at Ottawa, 1891–1911

come in the winter and freeze the seeds in solid ice, the winter wheat crop often fails, as happened in several of the years included in the reports (indicated by the broken top line in the graph). As in Figure A.1, the dashed lines indicate the mean value for each series.

In barleys, the distinction is between two-row barley, used mainly in brewing, and six-row, used mainly as animal feed. Two-row is a bit heavier (49.1 lb/bu here vs. 47.5 lb/bu for six-row), and this is true almost everywhere.

The ratio between the average of the two kinds of wheat and the average of the two kinds of barley for each year is shown on the bottom line. It is only possible to calculate this ratio for the years in which all four cultivars were grown; the average for the years in which such a comparison is possible is 0.798, which is virtually identical to the ratio $\frac{4}{5}$.

Behavior of specific varieties

While certainly helpful, the two sets of data just cited still don't represent the behavior of a specific variety of grain grown in a specific climate over an extended period of time. Such data turns out to be extremely difficult to find, for two reasons: first, because researchers are engaged in the continuous improvement of yields and have no reason to continue gathering data on older varieties once higher-yielding varieties have been identified; and second, because the companies that subsidize such research nowadays wish to keep research data private for competitive reasons. And whatever data might be relevant and (in theory) publicly available is hardly ever published, simply because there is little interest in it outside of the research community.

The longest more-or-less continuous record of the test weight of a particular variety of grain that I know of comes from the Cereal Laboratory in Svalöv, Sweden, one of the oldest such laboratories in the world. Missing just six years out of 87, the lab included in its research every year from 1892 to 1978 a planting of Sammet, a very old national variety of winter wheat from the time before plant breeding started. The test weight data for this period is shown in Figure A.3;[9] note that the weights are in kg/hl, not lb/bu.

The Sammet data demonstrates that even a small planting of wheat will, over the course of time, establish an average test weight for that kind of wheat in that location. A linear trend line for the 87 years

Appendix A. Grain weight as a natural standard

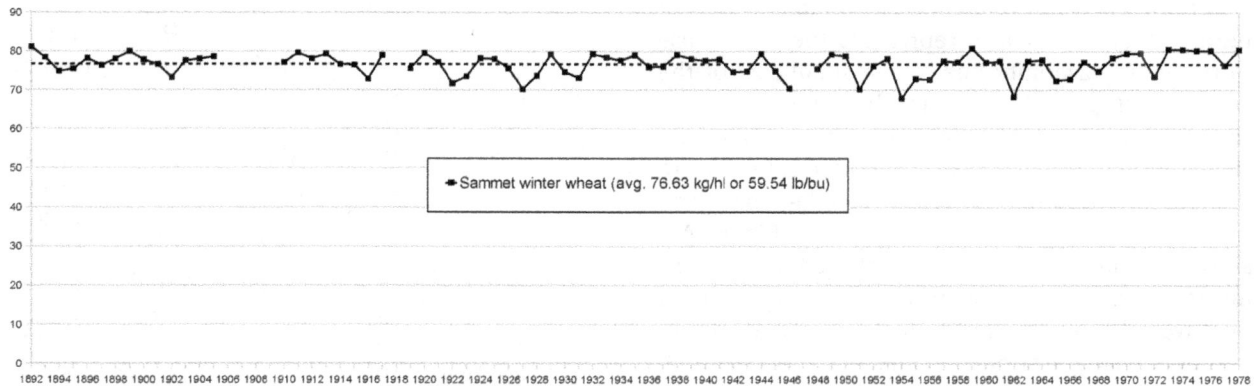

Figure A.3: Test weights of the old winter wheat variety "Sammet" at Svalöv, Sweden, 1892–1978

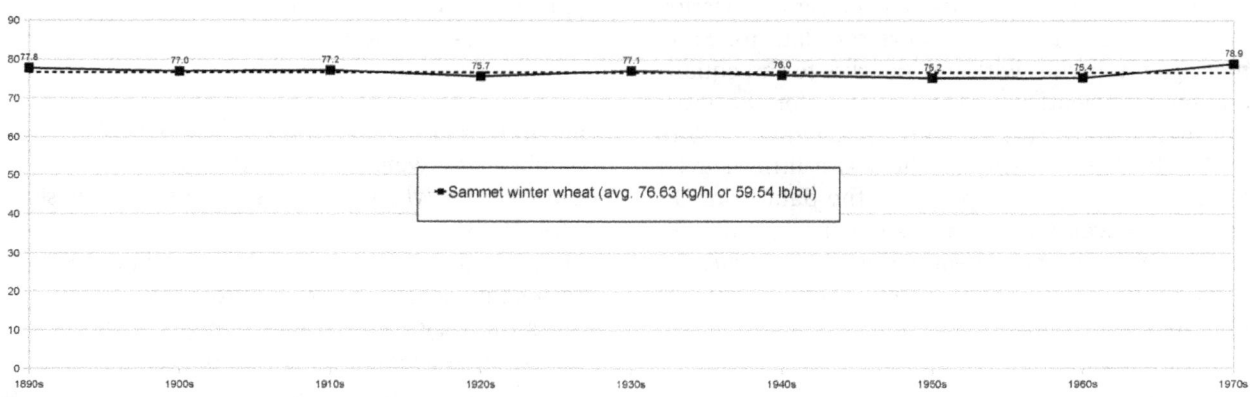

Figure A.4: The Sammet winter wheat data in 10-year intervals

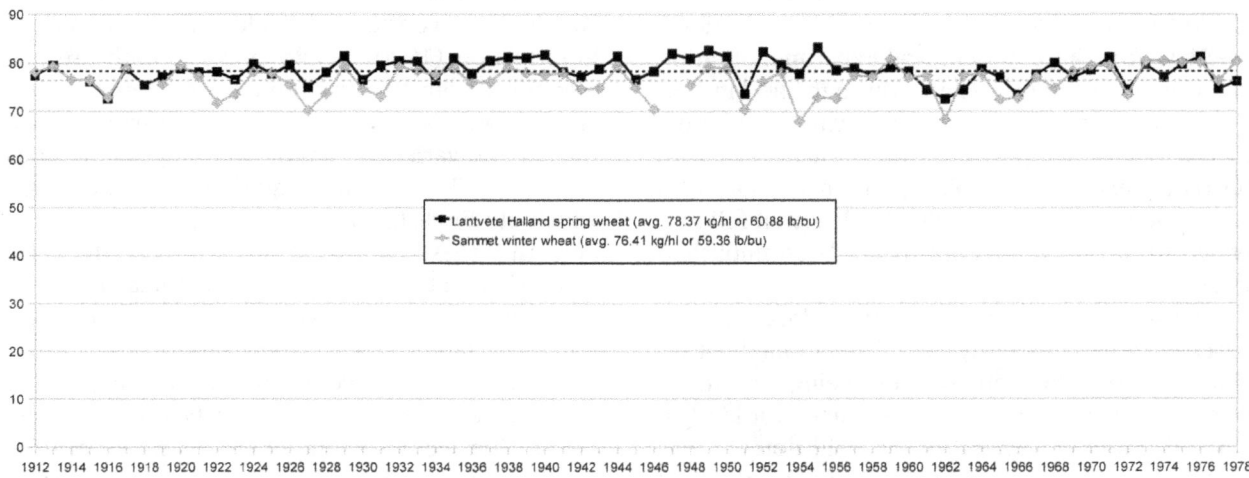

Figure A.5: Winter and spring wheat at Svalöv, Sweden, 1912–1978

Appendix A. Grain weight as a natural standard

is virtually identical to the mean data line displayed here. This data set also represents the worst case in terms of variability due to the small size of the sample (compare this with the behavior of the national averages in Figure A.1). But the extended period over which the data was gathered allows us to demonstrate another general principle—that even this variability smooths out over time. Figure A.4 shows the same data again, but plotted as the average for each decade. Any of the data sets presented here would show the same effect if they had continued over a longer period of time.

Another rare set of data from Svalöv shows that the relationship between the test weights of winter wheat and spring wheat is one of the parameters that actually does vary depending on location and climate. Figure A.5 shows the data for Sammet again together with test weights for an equally old spring wheat, Lantvete Halland, for the years during which the two were grown together, 1912 to 1978. In Sweden, spring wheat is a little heavier than winter wheat—opposite to the data from Ottawa. However, both data sets put the average test weight of traditional wheat varieties in roughly the same range: 59.3 to 61.3 lb/bu at Ottawa, 59.4 to 60.9 lb/bu at Svalöv.

The Cornell data

In 1985, I happened to be living close to one of the world's foremost agricultural research institutions, the New York State Agricultural Experiment Station at Cornell University. There, through the kind offices of Professor Mark Sorrells, I was given access to the raw field data for winter wheat and winter barley experimental plots going back almost 30 years and was able to compile the data that follows, which to my knowledge has never been published in any form.

Central to the usefulness of this data was the practice (now apparently long discontinued) of including among the many varieties being researched a few old wheat and barley cultivars as a kind of varietal museum. Based on the field notes, I was able to compile complete test weight, height, and yield results for 12 winter wheat varieties in an almost unbroken series for the 10 years 1969–1978

Variety	Year	Avg. Wt. (lb/bu)	Avg. Ht. (in.)	Avg. Yield (bu/ac)
Honor	1920	56.97	47.74	43.67
Forward	1920	59.97	45.72	46.06
Valprize	1930	58.42	45.91	41.07
Yorkwin	1936	57.87	46.72	46.42
Nured	1938	59.72	45.66	43.27
Cornell 595	1942	57.93	46.42	48.16
Genesee	1950	57.88	46.37	51.91
Avon	1959	57.85	43.80	51.25
Arrow	1971	58.08	38.67	52.37
Ticonderoga	1974	54.46	34.77	55.93
Houser	1977	55.27	35.93	64.92
Purcell	1979	57.33	37.64	66.51

Table A.1: Average characteristics of 12 varieties of Cornell winter wheat

(height measurements were not taken in 1970) and complete test weight and yield figures for the six oldest winter wheat varieties for the 20-year span 1959–1978.

Despite the one missing year, the data set with the height measurements is interesting because it shows the very considerable changes in the wheat plant itself as a result of modern breeding practices. Table A.1 gives the averages for the 10-year period 1969–1978 for the 12 oldest varieties, listed in the order in which each one was publicly introduced.[10]

The two oldest varieties on this list, Honor and Forward, are considerably older than their year of introduction (1920) would suggest; both are just selections from two traditional 19th century varieties, Dawson's Golden Chaff and Fulcaster. So like the two old varieties from Svalöv, they are about as close as we'll get to the traditional varieties upon which old English statutes were based. Note, however, that these figures do not represent the "best quality" wheat assumed by the statutes (more on that below) but rather average quality from a location never famous for its wheat production.

The results of modern breeding are evident in the changes in height and yield beginning in the 1950s. The average height of the last three varieties is 22 percent less than the average height of the first three, whereas the yield is 43 percent greater. (The two parameters are connected: reducing the height

Appendix A. Grain weight as a natural standard

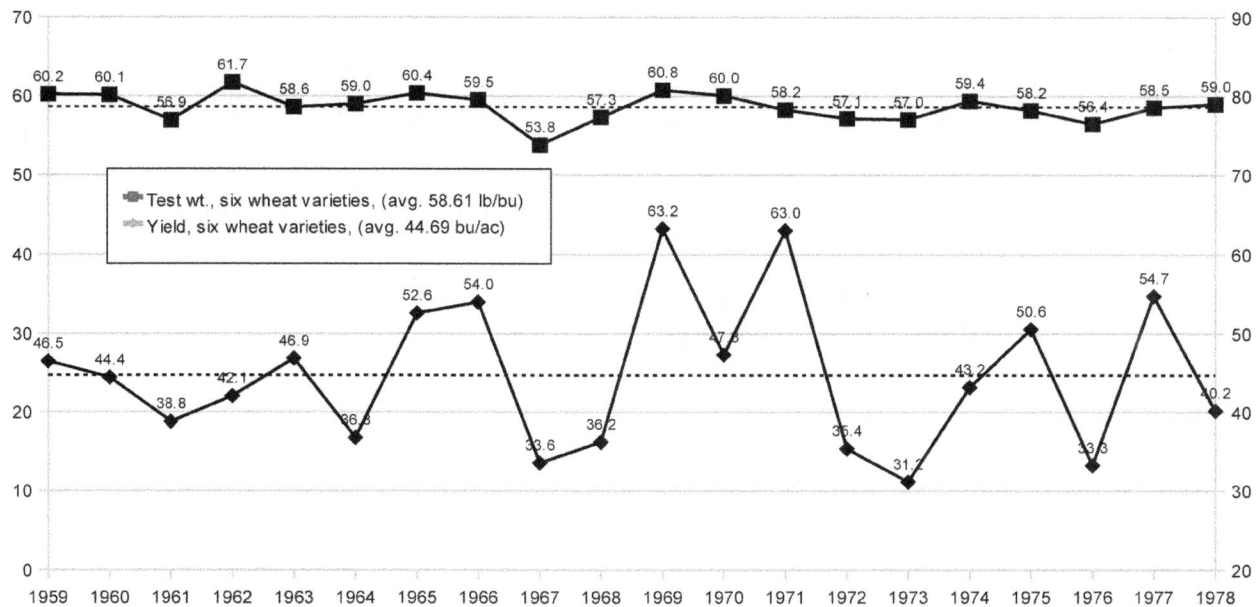

Figure A.6: Test weight vs. yield at Cornell, 1959–1978

of wheat plants reduces the amount lost by "lodging" when plants are blown over in severe weather.) Test weight, by contrast, changed much less over this period, though there is a slight trend downward, the three newest varieties averaging about 5 percent less in weight than the three oldest. Advances in breeding actually accelerated after 1979, and it can be seen from the early changes shown here that data from experiments later than the mid-1980s may not be the best for investigating the behavior of traditional wheat varieties.

Fortunately, of the seven oldest varieties listed above, six (the exception being Cornell 595) were represented in field notes with enough data to provide a complete 20-year series for test weights and yields. As with the Canadian and Swedish data presented earlier, the year-to-year averages are far more variable than seen in the U.S. national statistics we started with; instead of averaging an entire country's annual output, we're looking here at samples of rarely more than three rows per variety, sometimes two, and in many cases just one, yielding typically 1–3 pounds of seed to be weighed per row. The results are more variable than would have been experienced even by a single traditional farmer, and they can therefore, like the Canadian and Swedish experimental results, be considered a "worst case" with regard to variability.

Figure A.6 shows the average test weights of the six old winter wheat varieties on the top line and their average yields each year below. As before, the dashed line shows the 20-year average.[11] Variable as the results are from the minuscule samples grown in the tests, it's clear that test weight is far less variable than yield. (Yields are plotted against the axis on the right in order to move the line down so that it doesn't visually interfere with the test weight line, but the scale is basically the same for both sets of figures.) Even in disastrous years like 1967, 1968, 1972, 1973, and 1976, the test weight of the little wheat that was produced is usually within just a few percent of the weight of the wheat produced in the best ones. This is a key fact about the bulk weight of wheat.

Barley in New York gets less attention than wheat, but once again I was very fortunate to find continuous test weight data on two old varieties of winter barley for the same 20-year span: Wong, introduced in 1941, and Hudson, introduced in 1951. Figure A.7 shows the same test weight data for wheat seen in Figure A.6 on the top line, followed by test weights of the two old barley varieties, and then the

Appendix A. Grain weight as a natural standard

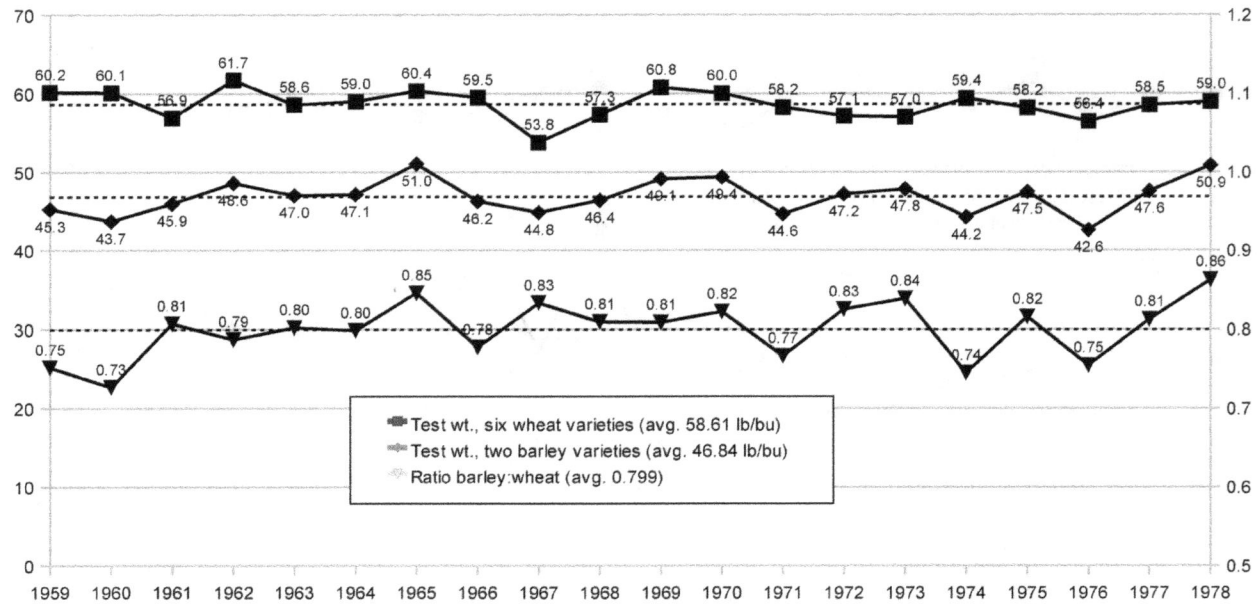

Figure A.7: Winter wheat and winter barley at Cornell, 1959–1978

ratio between these two numbers for each year.

This comparison of the same six wheat varieties with the same two barley varieties grown in the same location over a period of 20 years is probably as good as it gets in determining the relative weights of wheat and barley under experimental conditions, and the overall average of 0.799 for this ratio, taken with the 0.794 in the U.S. national data and the 0.798 in the Ottawa Experiment Station data, shows fairly conclusively that $\frac{4}{5}$ (0.8) is the simplest possible accurate figure for the ratio between the average bulk weights of traditional North American barley and traditional North American wheat. It is more accurate than either of the other two candidates ($\frac{3}{4}$ or 0.75 and $\frac{5}{6}$ or 0.833...) and simpler than any other reasonable alternative. And as noted in the text, this ratio is fixed in our legal definitions of bushels by weight (in most states and in Federal regulations, the weight bushel of barley is 48 pounds, and in all states the weight bushel of wheat is 60 pounds).

It should be recognized, however, that most of the barley in question is the six-row variety associated with classical grain production, which is nowadays used mainly for animal feed. For brewing, modern practice is to use two-row barley, which (as noted earlier) is a little heavier on average. In Sweden, for instance, the test weight of all wheat produced in the country for the 10-year period 1975–1984 (both winter wheat and spring wheat) averaged 79.48 kg/hl or 61.74 lb/bu, whereas the test weight of all barley produced in the country during that period (virtually all of which was two-row) averaged 69.19 kg/hl or 53.75 lb/bu. The barley:wheat ratio, therefore, was 0.87. This is an extreme example, but it does explain the $\frac{5}{6}$ barley:wheat ratio seen occasionally in legal grain weight equivalents for malting barley.

I should also note that the 64–80–100 density canon described in the text (barley weighs $\frac{4}{5}$ wheat, wheat weighs $\frac{4}{5}$ wine) is probably not appropriate to interpretation of the oldest systems, because the earliest civilizations were not yet based on wheat and wine. For the Mesopotamians and probably for the very early Egyptians, the relationship of most importance would have been between the weights of barley and water, barley being the staple food and the combination of barley and water providing the staple beverage, beer. The simplest workable barley:water ratio is $\frac{5}{8}$, and the Mesopotamian system of weights and measures lends itself readily to an analysis on this basis.[12]

Which wheat?

The old statute definitions of our bushel and gallon are based on weights of wheat. But given the dependence of test weight on quality, the question is: which grade of wheat did the authors of the statutes have in mind?

There doesn't appear to be much room for controversy on this. Both sets of statute definitions refer to grains of wheat "from the middle of the ear," and in practical terms (consider what would be involved in filling a bushel this way) the term must mean "hand picked" in the larger sense, that is to say, "wheat of the best quality."

Defining measures based on weights of grain makes it possible to carry out many sorts of practical calculations concerning the number of capacity measures of grain that a beast of burden can carry in comparison to the food needed to feed it, or the number of capacity measures of grain that a ship can carry if it is limited to displacing a certain volume of water, and so on. What's wanted here is not the weight of average wheat from the field, much of which is not worth transporting, and not the weight of the absolute best possible wheat, because that would be unrepresentative of trade, but rather *the average bulk weight of best-quality wheat.* In modern terms, this would be the average weight of export wheat, or wheat of the highest grade.

As just mentioned, the minimum allowable test weight for U.S. Number One wheat is 60 lb/bu; this goes back to a system of grades developed by the Chicago Board of Trade beginning in 1859.[13] Earlier definitions of grades in terms of test weights are not easy to find, but two stand out. In Paris in the mid-1700s, most experts put the test weight of the "elite" or best quality wheat at 250 livres per setier;[14] this is 60.90 lb/bu.[15] And in testimony before a Parliamentary Committee in 1834, the following benchmarks were given:[16]

In good seasons:

    Best quality wheat
        64 lb/Imp bu (62.00 lb/bu)
    Second quality wheat
        62 lb/Imp bu (60.06 lb/bu)
    Third quality wheat
        59 lb/Imp bu (57.16 lb/bu)

In bad seasons:

    Best quality wheat
        60 lb/Imp bu (58.13 lb/bu)
    Second quality wheat
        57 lb/Imp bu (55.22 lb/bu)
    Third quality wheat
        55 lb/Imp bu (53.28 lb/bu)

These early French and English estimates of the "the average weight of best quality wheat" put it somewhere in the range 61–62 lb/bu.[17]

Export wheat

Taking another angle, in 1985 I wrote to the trade embassies of a number of countries asking for recent test weight statistics for export-quality wheat. Nine of them replied with statistics full enough to be useful to this study; the results are listed in Table A.1 and give an average value for "export wheat" of 61.3 lb/bu.

The IWWPN

In a third approach to establishing "the average weight of best-quality wheat," I used data from the premier international study of wheat characteristics, the International Winter Wheat Performance Nursery conducted by the University of Nebraska with the cooperation of the U.S. Animal and Plant Health Inspection Service, the Food and Agriculture Organization of the United Nations, and the International Maize and Wheat Improvement Center. Each year for almost two decades, some 40 to 60 sites in two to three dozen different countries grew the same 30 varieties of winter wheat and then collated the results. The varieties grown each year changed slowly as the study progressed and the more productive varieties were selected, but some were kept in the mix as controls for extended periods of time.

The IWWPN results were published in annual volumes that collectively take up more than a foot of shelf space, but you will not find them in any bookstore or even online; information like this is kept within the research community. Figure A.8 shows

Appendix A. Grain weight as a natural standard

	Type of wheat	Test wt. (lb/bu)	Statistical basis
Australia (NSW)	Durum wheat	62.26	All exports, 1978/79 through 1984/85; total sample more than 49,406 tons[18]
Ireland	Winter wheat	60.18	Average of nine varieties in 1983/84[19]
	Spring wheat	60.85	Average of five varieties in 1983/84[20]
Italy	Bread wheat	63.22	National avg. 1980–1983 in study by Italian National Institute of Nutrition[21]
	Durum wheat	62.58	Avg. of durum processed by the Barilla company, 1980/81–1984/85 [22]
Jordan	Bread wheat	60.91	Ten-year averages (1975–1984) for two ordinary cultivars[23]
	Durum wheat	60.75	Ten-year averages (1975–1984) for four ordinary cultivars[24]
Norway	Bread wheat	59.75	Annual averages of 10–15,000 samples per year, 1977–1984[25]
Portugal	Bread wheat	62.49	Annual averages for 1981–1985; total sample 1,453,627 tons[26]
	Durum wheat	63.10	Annual averages for 1981–1985; total sample 91,424 tons[27]
Sweden	Winter wheat	62.28	Annual averages for 1975–1984 (entire harvest, both export and domestic)[28]
	Spring wheat	60.97	Annual averages for 1975–1984 (entire harvest, both export and domestic)[29]
West Germany	Bread wheat	58.59	Annual averages for 1977–1984[30]
U.S.A.	Bread wheat	60.98	All U.S. export wheat (except durum), 1982–1984[31]
	Durum wheat	60.53	All U.S. export durum, 1982–1984[32]
Average	All types	61.30	

Table A.1: Commercial wheat statistics (1985)

Appendix A. Grain weight as a natural standard

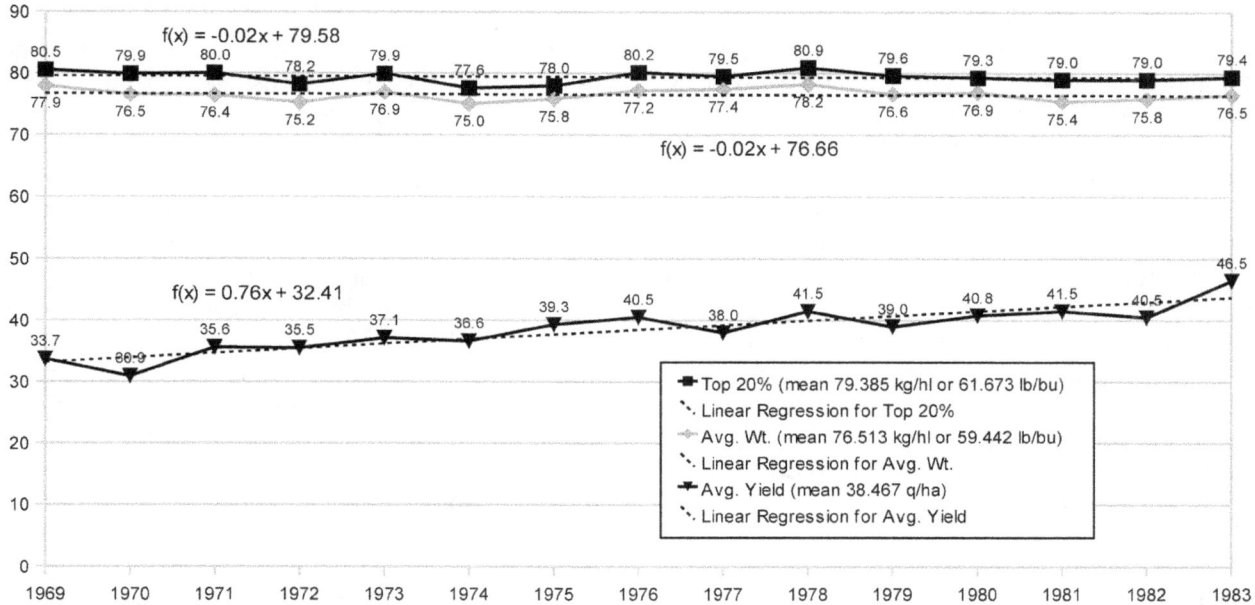

Figure A.8: Fifteen years of data from the International Winter Wheat Performance Nursery

test weight and yield data from the first 15 years of the study (1969–1983).[33]

The top two lines are the average test weights for all sites, as reported in kilograms per hectoliter.

The lower of the two lines is the average for all 30 varieties at all the sites. The sites spanned a wide range of altitudes and climates, including several locations not well suited for growing winter wheat and some where doing so is barely possible, so the data summed up in the overall average includes a lot of failed wheat that was barely testable. This mix, therefore, represents wheat that is decidedly poorer in average quality than "average best quality wheat." Over the 15 years of the experiments reported here, that average for all varieties at all sites was 76.513 kg/hl or 59.442 lb/bu.

To simulate "best quality wheat" based on this study, I selected the six varieties that were ranked as the best in test weight averaged over all the sites each year—in other words, the top 20 percent (six out of 30) each year. The results are shown in the top line of the chart. The top 20 percent averages 79.385 kg/hl or 61.673 lb/bu.

As in the previous examples, the data over a number of years demonstrates how steady average grain weights are. In the IWWPN study, it's particularly notable that the average test weight of the shifting assortment of varieties remained virtually static even as aggressive breeding during this period caused a substantial increase in productivity, as shown by the yield line toward the bottom of the graph; the dashed line shows the climb upward in productivity from less than 35 quintals per hectare at the beginning of the 15-year period to close to 45 at the end.

The dashed lines for the test weights are also linear trend lines, unlike the simple mean value lines shown in previous charts, and while the test weights remain so steady that it's hard to tell the difference, the trend lines show a slight decrease in test weight due to breeding for other characteristics; as indicated by the formulas in the chart, both the top 20 percent and the overall average show a downward slope of –0.02 over the 15-year period. Nonetheless, the fact that test weights remained relatively stable despite large-scale genetic changes over the period shown is an impressive demonstration of the stability of this natural constant.

The thing to bear in mind when considering any of this data with regard to traditional practices is that the longer the time interval used for each data point, the less variation in the average there will

Appendix A. Grain weight as a natural standard

be. For example, if the national data in Figure A.1 had been gathered over 300 years instead of 30 and each data point represented a 10-year average instead of just one, the test weights would show as virtually a flat line. I'm not saying that the ancients necessarily kept statistics quite like this (though they certainly could have), but after a few centuries of growing the same grain in the same place, they must have had a pretty good idea of what it weighed on average, and once that average was fixed in place by its realization in standard weight and capacity measures, it would provide a sound basis on which to calculate the answers to a large number of practical problems relating to the transportation and storage of trade goods.

Conclusions from the statute definitions

To sum up the data given above for "best quality wheat" to the nearest tenth of a pound:

Elite wheat (Paris, 18th century):
60.9 lb/bu (198.2 grains per in^3)

Best wheat in a good year (London, 1834):
62.0 lb/bu (201.8 grains per in^3)

Export wheat, nine countries (1985):
61.3 lb/bu (199.5 grains per in^3)

IWWPN top 20 percent (1969–1983):
61.7 lb/bu (200.8 grains per in^3)

Based on the data, we can say with reasonable assurance that the most likely value for "average best quality wheat" will be around 61.5 lb/bu or 200.2 troy grains per cubic inch, and it seems safe to say that no value below 58.1 lb/bu (given in 1834 as the weight of the best wheat in a bad season) could represent anyone's notion of wheat of the best quality. In fact, it appears unlikely that "wheat of the best quality" used as the basis for a system of measures would have a weight less than the minimum for ordinary commercial U.S. No. 1 wheat, 60 lb/bu.

These considerations demonstrate conclusively that the Winchester bushel cannot be the measure intended by the definition of a bushel as 64 troy pounds of wheat (i.e., 8 gallons each holding 8 troy pounds of wheat). Wheat that weighed 64 troy pounds (64×5760 = 368,640 grains) per Winchester bushel—in other words, 52.7 lb/bu—would not even be sold for human consumption in this country.

If, on the other hand, the Winchester bushel holds 64 tower merchant's pounds of wheat (64×6750 = 432,000 grains), as it was apparently defined in the 13th century statutes, then that wheat weighs 61.7 lb/bu, and if the wine gallon holds 8 troy pounds of wheat, as it was very specifically defined in 1496, then that wheat weighs 61.3 lb/bu. These figures nicely bracket the most likely value for "average best quality wheat" found above. The test weight data also lends credence to the theory that the ale gallon of 282 in^3 was intended to hold 8 avoirdupois pounds of wheat; the wheat in that case would weigh 61.0 lb/bu, and if we go by the oldest example of this standard, the ale quart of Elizabeth I (70.22 in^3), the wheat would weigh 61.2 lb/bu.

The imaginary wine bushels of Henry VII

Based on the physical evidence, it appears that the statute 12 Henry 7 c. 5 (1496) created a phantom set of grain measures based on troy weight—much to the confusion of later authorities and modern historians.

The measure of 8 wine gallons certainly existed; this is the firkin defined by the statute 23 Elizabeth 1 c. 8 §4 (1581) quoted in Chapter 3 as a subdivision of the barrel of 32 wine gallons, a measure that can be traced back several thousand years. But there appears to be no evidence whatever that this measure of 8 wine gallons was actually used as a bushel for measuring grain and other dry substances. For example, there is, as far as I know, no exchequer or city standard bushel measure (meaning a wide vessel with handles, like the one on the back cover of this book) based on the wine gallon standard.

Nevertheless, editors in the centuries after the statute of 1496 faithfully carried forward this mythical standard for grain. For example, a commonplace book dated 1682 and dedicated to James, Duke of York, contains the following:

Wyne, Oyle and Hony Measures

One Gallon conteynes 8 pound Troy, by which the vessels following are measured:

1 Tunne conteynes: 2 Pipes or Butts, 3 Punchions, 4 Hogsheads, 6 Tierces, 8 Barrells, 14 Rundlets, 252 Gallons....

Nothing could be clearer, if you understand that the pounds are pounds of wheat per 12 Henry 7 c. 5. But then it goes on to say

Grayne

1 Last conteynes: 10 Quarters, 20 Cornookes, 40 Strikes, 80 Bushels, 320 Pecks, 640 Gallons, 5120 Pints. Measured by Troy weight, 8 pound to a gallon.[34]

Taken together, these statements mean that a wine gallon is identical in capacity to a Winchester gallon—which no one who worked with the units could possibly have believed; wine and Winchester standards were both provided to the cities and towns, and the gallons of each were of obviously different sizes.

Later examples confirm that the part about the Winchester gallon containing 8 troy pounds of wheat had become a meaningless formula, like the one about a penny weighing the same as 32 wheat corns. An arithmetic published in 1658 as much as says this:

> At London, and so all England thorow, are used two kind of Weights and Measures, as the *Troy* Weight and the *Haberdepoise* [by this time, *haberdepoise* means modern avoirdupois]. From the *Troy* Weight is derived the proportion and quantity of all kind of dry and liquid Measures, as Pecks, Bushels, Quarters, &c. Wherewith is bought and sold all kind of Grain and other Commodities mete by the Bushel. And in liquid, Ale, Beer[,] Wine, Oyl, Butter, Hony, &c. Upon these Grounds and Statutes is Bread made, and sold by the *Troy* weight: and so is Gold, Silver, Pearl, precious Stones, and Jewels. The least quantity of this *Troy* Weight is a Grain: twenty four of these Grains make a penny weight, twenty penny-weights an Ounce, and twelve Ounces a pound[,] two pound or pints of this weight maketh a quart.

Appendix A. Grain weight as a natural standard

> And so ascending into bigger quantities, is produced the Measures whereby are sold our other natural sustenance: *viz.* Ale or Bær, with all other necessary commodities, as Butter[,] Hony, Herrings, Eels, Sope, &c. All which last before rehearsed, though their Measures (wherein they are contained) be framed and derived from the *Troy* weight, yet they are in traffique with divers Commodities, as Lead, Tin, Flax, Wax, with all other Commodities, both of this Realm, and of other Forreign Countries whatsoever, bought and sold by the *Haberdepoise* weight after sixteen Ounces to the pound, and 112 pound to the hundred Weight.[35]

In other words, the capacity measures were acknowledged to be defined by the troy-based definitions in 12 Henry 7 c. 5 (1496), but in practice, troy weight was no longer used in connection with those measures, and thus there would never be occasion to discover that the Winchester bushel and the other measures derived from it bore no relation to troy weight.

A comparison of summaries regarding the system of weights and measures in various 18th and 19th century reference works shows the extent to which the editors of these works slavishly copied one another. Here is a passage from a 1753 compendium for tradesmen and shopkeepers:

> The most common weights used throughout this kingdom are two: the troy weight, and the avoirdupois weight. The troy weight contains 24 grains in a penny weight, 20 dwts. to an ounce, and 12 ounces to a pound; and is used only in weighing bread, gold, silver; and by the apothecaries in their medicines; 8 lb. troy is a gallon, 16 lb. a peck, and 64 lb. a bushel; and hereby Weight and Measure are reduced into one another.
>
> Wet Measure is also derived from this pound troy, both on land and on shipboard, as also grain and corn as before named; for first, these 12 ounces made into a concave measure is named a pint, 8 of these pints make a gallon (containing 231 cubical inches) of wine, brandy, cyder, &c. according to the standard of the exchequer. From hence is also drawn

Appendix A. Grain weight as a natural standard

the assize measure of all vendible casks: a hogshead is to contain 63 gallons, a tierce 42 gallons, a pipe 126 gallons, and a ton 252 gallons, and weighs 1890 lb. averdupois, or 2016 lb. troy.[36]

In a standard 18th century reference book on commercial law that went through many editions over the course of several decades we find the following (sixth edition, 1773):

> The Weights in common Use throughout *Great-Britain,* are Troy and Avoirdupois; the former consisting of Grains, Pennyweights, Ounces, and Pounds, whereof 24 Grains make a Pennyweight, 20 Pennyweights an Ounce, and 12 Ounces a Pound, by which Bread (in Corporation Towns only) Gold, Silver, and Apothecaries Medicines are weighed; and to this Weight Corn Measures are reducible, as 8 lb. Troy makes a Gallon, 16 lb. a Peck, and consequently 64 lb. a Bushel; Liquid Measures are also dependant on it, as their Concavities correspond in their different Sizes thereto, from a Pint consisting of 12 Ounces (or a Pound) up to a Ton, containing 252 Gallons, and weighing 2016lb. [Troy] or 1890 lb. Avoirdupois; 2 Pints make a Quart, 4 Quarts a Gallon (containing 231 Cubical Inches) 63 Gallons a Hogshead, 42 a Tierce, 126 a Pipe, and 252 a Ton of Brandy, Cyder, Wine, &c.[37]

Compare this with the following passage (legal footnotes omitted) embedded in a long discussion of statutes relating to weights and measures published half a century later in an 1824 treatise on the laws of commerce:

> The legal *weights* in common use throughout Great Britain, are troy and avoirdupois; the former consisting of grains, pennyweights, ounces, and pounds, whereof twenty-four grains make a pennyweight, twenty pennyweights an ounce, and twelve ounces a pound, by which bread, gold, silver, and apothecaries medicines are weighed; and to this weight corn measures are reducible, as 8lbs. troy make a gallon, 16lbs. a peck, and consequently 64lbs. a bushel.... Liquid *measures* are also dependent on this weight, as their concavities correspond in their different sizes thereto, from a pint, consisting of 12 ounces or a pound, up to a ton, containing 252 gallons, and weighing 2016lbs. [troy], or 1890 avoirdupois; two pints make a quart, four quarts a gallon, containing 231 cubical inches; 63 gallons a hogshead, 42 a tierce, 126 a pipe, and 252 a ton of brandy, cider, wine, &c.[38]

With the exception of two sentences in the middle referencing 12 Henry 7 c. 5 and 13 & 14 William 3 c. 5, which I've left out here, this passage is copied almost verbatim from the one above it ("as their concavities correspond" indeed). But my point in quoting these various accounts is not the cavalier treatment of intellectual property before the advent of modern copyright laws but rather the astonishing lack of attention in all of these sources to the logical consistency of what's being said, as demonstrated by the spectacularly bad conversion between troy and avoirdupois weight in these examples.

If 2016 pounds troy equals 1890 pounds avoirdupois (as clearly stated in all three of the passages quoted above), then either the avoirdupois pound weighs 6144 troy grains instead of 7000 or the troy pound weighs $6562\frac{1}{2}$ grains instead of 5760. A quantity of 252 wine gallons could weigh 1890 pounds avoirdupois (as stated in all three references) only if liquids weighed about ten percent less than they actually do. All three authors state, correctly, that the gallon of which 252 make a tun is the wine gallon of 231 in^3; if this gallon were a Winchester gallon instead, then a 252-gallon weight of 1890 pounds avoirdupois would imply wheat weighing 60 lb/bu, which is exactly the legal weight of wheat in the U.S.; and this might explain, by some very confused process of reasoning, where the figure of 1890 pounds avoirdupois came from. But nothing can explain the stated equivalence of 1890 pounds avoirdupois and 2016 pounds troy.

This is no trifling discrepancy; the conversion is off by about 14 percent. Someone, somewhere along the line, multiplied 8 troy pounds times 252 gallons to get 2016 troy pounds of wheat per wine tun (which has never been a measure for grain), and someone—probably someone else—divided 60 avoirdupois pounds per Winchester bushel of wheat by 8 to get $7\frac{1}{2}$ avoirdupois pounds per gallon and

Appendix A. Grain weight as a natural standard

therefore 1890 avoirdupois pounds per tun of 252 *Winchester* gallons (which has never existed as a measure at all). And no one seems to have considered whether the two definitions were consistent with each other.

It's possible that further research into old manuscripts may someday unravel how the confusion became embedded in these figures. The thing that's perfectly clear at this point is that generations of authorities on the subject cannot possibly have understood what they were passing down through the years. And this includes the repeated references to a dry bushel with a capacity of 8 wine gallons, which appears never to have actually existed.

As late as 1864, we find in the 11th edition of *The Shipmaster's Assistant,* a standard reference book for American ship captains, the following passage (pp. 362–63):

> 8 lbs. Troy are considered as equal to a gallon; 16 lbs. equal to a peck; and 64 lbs. to a bushel, where weight and measure are reduced into each other. . . .
>
> Liquid measure is derived from the pound Troy, as well as the measure of grain; 12 ounces or a pound being considered as equal to a pint, and 8 pints equal to a gallon (of 231 cubic inches) of wine, brandy, cider, &c., of which the standard is preserved in the state department at Washington.

It must be assumed that this passage was simply ignored by all real ship captains, who from the time of Elizabeth worked mainly in avoirdupois pounds and by the time this was published knew perfectly well that a canonical U.S. bushel of wheat by weight was actually set by law in all the states at 60 avoirdupois pounds, and that this was also, by that time, the minimum acceptable test weight for Number 1 wheat at the Chicago Board of Trade. Sixty pounds avoirdupois is about 73 troy pounds to the bushel, not the 64 troy pounds repeated here as definitional.

The discovery of actual bushel measures of 8 wine gallons, if that ever happens, would certainly change our picture of the implementation of the statute 12 Henry 7 c. 5; in that case, it would more resemble the short-lived attempts to define newfangled bushels in 19th century New York and Ohio, because if such an attempt had been made, it must have been a complete failure. But in any case, it's clear that the 64 troy pound definition doesn't fit the Winchester bushel, whereas the definition in terms of the tower merchant's pound in the *Tractatus* fits it perfectly.

"A pint's a pound"

For about a century and a half in the UK, children were taught the mnemonic "a pint of pure water weighs a pound and a quarter," which was true by definition in the Imperial system. But here in the U.S. we kept an even older saying, "a pint's a pound the world around."

The proverb doesn't actually apply very accurately to the U.S. pint and the avoirdupois pound. As noted earlier, a wine gallon of water weighs $8\frac{1}{3}$ pounds, so in reality the U.S. pint of water weighs $\frac{25}{24}$ or 1.041666... avoirdupois pounds.

Some have tried to explain this discrepancy by representing the adage as merely an aid in remembering that both pint and pound are divided into 16 (different) ounces.[39] But the preceding analysis of wheat weights allows us to demonstrate that, in fact, the saying applies literally and accurately to all four of the old English capacity standards if we understand which pound goes with each one.

Table A.2 shows these standards, expressed here in terms of their pints, and their probable bases in the weights of wheat and wine. In the case of the wine pint, the definition is given to us by the statute of 1496; the interpretations of the rest line up perfectly with a common estimation of the weight of good quality wheat and a shared canon of density ratios (wheat weighs $\frac{4}{5}$ as much as wine).

The pints are listed in their presumed order of age; we are probably safe in assuming (at least) that the standards based on troy and tower weight came before the ones based on avoirdupois. In the case of the ale pint, we can choose between a later legal equivalence of 282 in^3 or go directly to the Exchequer standard of Elizabeth I. If the former, then the wheat weight implied by the average of the four standards is 199.74 troy grains per cubic inch or 61.36 pounds per bushel; if the latter, 199.94 grains per cubic inch or 61.42 pounds per bushel.

Appendix A. Grain weight as a natural standard

Old English pint standard	Official definition (cubic inches)	Interpretation	Implied weight of wheat (troy grains per cubic inch)	Implied weight of wheat (pounds per bushel)
wine pint	28.875 ($\frac{1}{8}$ gallon of 231)	1 troy pound (5760 grains) of wheat	199.481	61.281
Winchester pint	33.6003125 ($\frac{1}{64}$ bushel of 2150.42)	1 tower merchant's pound (6750 grains) of wheat	200.891	61.714
ale pint	35.11 ($\frac{1}{2}$ Exchequer ale quart of Elizabeth I)	1 avoirdupois pound (7000 grains) of wheat	199.373	61.248
	35.25 (later definition, $\frac{1}{8}$ gallon of 282)	1 avoirdupois pound (7000 grains) of wheat	198.582	61.005
Guildhall pint	28 ($\frac{1}{8}$ gallon of 224)	1 avoirdupois pound (7000 grains) of wine ($\frac{5}{4}$ wheat)	200	61.441

Table A.2: Weight bases of the old English capacity standards

Notes

[1] Hunt, *The Cereals in America*, 42.

[2] *Test Weight and Your Next Wheat Crop: A Fact Sheet.* Kansas State University, n.d. http://www.kscrop.org/resources/SeedManagement.pdf

[3] ADAMS, 193.

[4] *Public Statute Laws of the State of Connecticut*, bk. 1, 683.

[5] *Minutes of Evidence* (1862), 2.

[6] USDA *Grain Inspection Manual*, Instruction No. 918(GR)–6, 21.

[7] USDA *Weather, Crops, and Markets*, Vol. 1 No. 11 (March 1922) through Vol. 8 No. 11 (November 1931).

[8] See bibliography under "Canadian Parliament."

[9] T. H:l-Gumaelius, Svenska Cereallaboratoriet AB; personal communication, 20 December 1985.

[10] Before their introduction as named varieties, Arrow, Ticonderoga, Houser, and Purcell were recorded in the field data by code numbers. A table very similar to this showing height and yield averages for these same Cornell wheat varieties over the 10-year period 1966–1975 appears in Jensen, "Limits to Growth in Global Food Production," 317–20.

[11] I should note here for people who wish to get deeply into this that "bu" means two different things in these statistics. In test weights, as stated in the text, "lb/bu" means avoirdupois pounds of 7000 troy grains per Winchester bushel of 2150.42 in^3. In yields, "bu/ac" means units of 60 pounds (not Winchester bushels) produced per acre of land. In the case of the experiments summarized here, yields were actually calculated by weighing the seeds in grams and then calculating bu/ac using conversion factors based on the number of rows harvested in the test.

[12] In the Sumerian and Old Babylonian periods (roughly the third and second millennia B.C.), the system of smaller weights was talent = 60 minas, mina = 60 shekels, and shekel = 180 grains (Mesopotamian grains, not troy grains). The system of capacity measures for barley was bariga = 60 sila, sila = 60 (capacity) shekels, and shekel = 180 (capacity) grains (I've left out a measure called the ban of 10 sila to make the comparison easier). See Neugebauer and Sachs, *Mathematical Cuneiform Texts*, 4 and 6, and Robson, "Mesopotamian Mathematics," 70–71.

The most common value for the capacity of the sila found in ancient tablets is 21,600 (6,0,0) sila to the volume-SAR, the latter being a rectangular solid measuring 12 cubits by 12 cubits by 1 cubit, i.e., 144 cubic cubits (Robson, "Mesopotamian Mathematics," 123); this means that the sila was $\frac{1}{150}$ of a cubic cubit, that is (taking the cubit as 49.5 cm), 808.5825 mL or 49.3427 in^3. As mentioned in Chapter 2, two surviving representatives of this standard imply a unit averaging 808.435 mL (see pages 62 and 66).

The parallelism of the weight and capacity units, and the use of the same word (*gín*) for the weight shekel and the capacity shekel—like our use of the word *ounce* for both weight and capacity units—strongly suggests a connection between the two systems, and the weight of barley provides the link, just as the weight of wheat does in our own tradition. If barley canonically weighs $\frac{5}{8}$ as much as water, and the weight of water is reckoned according to the definition of Imperial measure (see page 65), then barley weighs 48.44 lb/bu and the sila of barley ($\frac{1}{150}$ cubic cubit) weighs 7780.3 troy grains. This implies a shekel of 129.67 troy grains (8.403 grams), which accords very well with the value of the Sumerian/Babylonian shekel established through various studies of the archaeological data. Thus the basic early Mesopotamian system was simply

1 capacity-shekel of barley weighs 1 shekel
1 sila of barley weighs 1 mina
1 bariga of barley weighs 1 talent

This interpretation fits the smallest units of weight and volume as well. We know that the Mesopotamian cubit was divided into 30 digits, so if the sila is $\frac{1}{150}$ cubic cubit, it has a volume of 180 cubic digits, and therefore each cubic digit of barley has a volume of 60 capacity-grains and weighs 60 weight-grains or $\frac{1}{3}$ shekel. Using the weight of barley above, the cubic digit of barley would weigh 43.224 troy grains; again, this is a very accurate value for $\frac{1}{3}$ of the actual shekel, and it corresponds to $\frac{1}{10}$ of the unit of 432 troy grains that pops up a couple of times in Chapter 2. This unit eventually became the weight of the Spanish coin called the real and later the Maria Theresa thaler, which remains in circulation to this day in the Middle East at a weight of 28.0668 grams or 433.14 troy grains.

It seems unlikely that I am the first person to observe these connections based on the barley:water ratio, but I can't recall seeing them noted elsewhere. A number of interesting relationships follow; for example, on this basis the cubic Mesopotamian foot of barley weighs 24 Egyptian deben or 60 troy pounds. As noted earlier, the volume of this cubic foot was, quite precisely, the legal bushel in the state of Connecticut until 1850 (see page 58).

[13] Hill, *Grain Grades and Standards*, 16.

[14] Kaplan, *Provisioning Paris*, 52–53.

[15] Prior to the introduction of the metric system, the French livre was 489.505847 grams and the setier of Paris for grain and other dry substances (except coal, salt, and oats) was 156.099639 liters (MARTINI, 413).

[16] *Report from the Select Committee on the Sale of Corn*, 61–62 (testimony of Mr. Patrick Stead, Corn-merchant and maltster). Conversions from pounds per Imperial bushel (lb/Imp bu) to pounds per U.S. bushel (lb/bu) are calculated based on the value for the Imperial bushel given by Order in Council (1889) as 2219.704 in^3 (see note on page 65).

[17] It should be noted that many of these early grading systems put the cutoff for *unmarketable* wheat much lower than we do today; for example, some 18th century French authorities put the test weight for the lowest grade of wheat as low as 220 livres per setier, which is 53.6 lb/bu, and the original Chicago Board of Trade grading system put the "Rejected" grade at 40 lb/bu. The consistency in wheat weights demonstrated in this appendix is at the upper end of the quality scale, not the lower.

[18] D. Heinrich, Scientific Officer, Australian Wheat Board. Personal communication, 10 October 1985.

[19] Noel Purcell-O'Byrne, Deputy Consul General, Consulate General of Ireland (New York City). Personal communication, 18 October 1985.

[20] Ibid.

[21] Richard B. Helm, Acting Agricultural Counselor, U.S. Embassy, Rome. Personal communication, 8 November 1985.

[22] Ibid.

[23] Kamal Hasa, Counselor, Embassy of the Hashemite Kingdom of Jordan. Personal communication, 18 November 1985.

[24] Ibid.

NOTES

[25] Kjell M. Fjell, Laboratory Manager, Statens Kornforretning (Norwegian Grain Corporation). Personal communications, 23 October and 5 December 1985.

[26] Rui Pascuela, Technical Director, Empresa Pública de Abastecimento de Cereals (EPAC). Personal communication, 13 January 1986.

[27] Ibid.

[28] T. H:l-Gumælius, Svenska Cereallaboratoriet AB. Personal communication, 7 November 1985.

[29] Ibid.

[30] Dr. Hoffman, Bundesminister für Ernährung, Landwirtschaft und Forsten. Personal communication, 29 October 1985.

[31] John W. Marshall, Director, Field Management Division, Federal Grain Inspection Service, USDA. Personal communication, 5 August 1985.

[32] Ibid.

[33] Research Bulletins Nos. 245 (July 1971) through 305 (March 1984) of the Nebraska Agricultural Research Station.

[34] Hall and Nicholas, *Tracts and Table Books,* 29.

[35] Ray, *Tap's Arithmetick,* 322–23.

[36] *General Shop Book,* s.v. "Weights and Measures of Great Britain."

[37] Beawes, *Lex Mercatoria Rediviva,* 728.

[38] Chitty, *Treatise on the Laws of Commerce and Manufactures,* 2:169–70.

[39] Editions of the *Encyclopædia Britannica* beginning with the first (1771) and continuing at least as late as the third (1797) dealt with the discrepancy simply by denying that it existed, stating that "the wine-gallon contains 231 cubic inches, and holds eight pounds averdupois of pure water" (s.v. "Gallon"). The difference between 8 pounds and $8\frac{1}{3}$ pounds being orders of magnitude greater than the precision of measuring instruments even in the 18th century, it's safe to assume that the people who wrote this never actually tried weighing a gallon of water but were relying on the proverbial "pint's a pound."

Appendix B. The English statute basis of U.S. weights and measures

Virtually all of the old English statutes were swept away in Britain with the passage of the reform of 1825 that created Imperial measure. But they still provide the legal basis for weights and measures in the U.S. wherever they haven't been replaced by state laws. This was explicitly recognized in some early state laws, beginning with the citation of 12 Henry 7 c. 5 (1496) in a 1661 Virginia statute noted in Chapter 3.

The English statutes in U.S. common law

The basic legal principle applying here is that the laws in effect in the colonies in 1776 were English laws, and upon independence, those laws stayed in effect until they were replaced by state or federal laws. Under the U.S. Constitution, Congress has the authority to "fix the standard of weights and measures," but with few exceptions (the most important being the legalization of metric units in 1866) it never actually has; it has just recognized, maintained, and distributed copies of standards already in existence. Consequently, the basic laws regarding weights and measures in the U.S. are state laws. And where not supplanted by later legislation, the laws are what they were when independence was declared.

Thus, in Thompson v. District of Columbia, 21 AD. C. 395 (1904), it was found that "... [I]n the absence of a standard fixed by Congress, the English standard, which was brought to the colonies and governed therein as part of the common law, and has been recognized by Congressional legislation, is the standard with which they [the Commissioners of the District of Columbia] are given the power to enforce conformity."[1] In United States v. Weber, 3 Ct. Cust. A19 (1912), it was held that "in the absence of any specific declaration by Congress as to the contents of a bushel of apples or the like, it will be presumed that a bushel is the bushel of English law and custom in 1776; and a bushel of apples is not a struck Winchester bushel, but that measure heaped."[2] And in Dwight and Lloyd Sintering Co. v. American Ore Reclamation Co., 263 F. 315 (1920) it was determined that "the standard weights and measures of England were brought to the colonies and became part of the common law of the land."[3] So it appears that old English laws relating to weights and measures are in theory still law in the U.S. wherever they have not been superseded by state law.

The part of the inherited English statute law that's not covered by state law will, of course, vary from state to state. I have found two official codifications of the old statutes as they were held to still apply to U.S. state laws, one from Kentucky in 1810,[4] and one from Maryland in 1811, reaffirmed in 1870.[5]

The 1810 Kentucky compilation listed the following English statutes as applicable in whole or part to Kentucky state law relating to weights and measures.

"stat. Incerti Temporis" (of uncertain time)[6]

2 Henry 6 c. 11 (1423)[7]

9 Henry 6 c. 8 (1430)

1 Richard 3 c. 13 (1483)

11 Henry 7 c. 5 (1496)[8]

23 Henry 8 c. 4 (1531)

The 1811 Maryland compilation (reaffirmed in 1870) listed the following English statutes as applicable in whole or part to Maryland state law relating to weights and measures.

Appendix B. The English statute basis of U.S. weights and measures

    Magna Charta [*sic*], 9 Henry 3 (1225)

    33 Edward 1 stat. 6 (1301)

    14 Edward 3 stat. 1 c. 12 (1340)

    25 Edward 3 stat. 5 c. 10 (1350)

    13 Richard 2 stat. 1 c. 9 (1389)

    11 Henry 6 c. 8 (1433)

    12 Henry 7 c. 5 (1496)

What strikes one immediately about these lists is their almost complete disjunction. With the exception of 12 Henry 7 c. 5, which defined troy weight and the wine gallon, the legislatures of Kentucky and Maryland arrived at two different sets of English statutes as still applicable to their weights and measures. While the general principle is clear, therefore, it's also apparent that a determination of which of the old laws still apply in areas that aren't specifically addressed by state law is a matter of some interpretation.

In fact, the two lists above are far from exhaustive.

English statutes relevant to standards of weight and measure

In the tables that follow, I've listed all the English statutes I could find relating to standards of weight and measure before U.S. independence in 1776. These are just the ones relating to the size of the standards, not to their administration, distribution, or enforcement. I hope in some future edition of this book to provide the complete texts of the relevant portions of these statutes as well.

The references in the right-hand column in each table give volume and page numbers for the three sources I consulted for this survey. "R" is Robertson's *Laws of the Kings of England from Edmund to Henry I* (1925). "G" is the official compilation authorized by George III and issued in 11 volumes from 1810 to 1828 under the title *Statutes of the Realm;* this is available in reprint editions and to scholars through online subscription services, and it has generally been taken as the authoritative source. "P" is Pickering's *Statutes at Large,* published from 1762 to 1775 (in the part that's relevant here), which was used to fill out a few years that aren't covered in the George III edition. Where Pickering's numbering differs from G, or where G itself notes a departure from earlier editions, I have recorded the difference; this is important because many earlier histories use the older numbering.

Appendix B. The English statute basis of U.S. weights and measures

Statute	Primary subject of relevant portion	Reference
3 Edgar c. 8, ca. 960	One system of measurement and one system of weights, such as used at London and Winchester	R 29
9 Henry 3, 1225 (*Articuli Magne Carte Libertatum*, "Magna Carta")	One measure of wine, one measure of ale, one measure of grain; widths of cloth	G I Charters 11
51 Henry 3, 1266[9] (*Assisa Panis et Cervisie*)	Grains of wheat; sterling penny; (tower) ounce and pound; (Winchester) gallon, bushel, and quarter	G I Statutes 200–01
31 Edward 1, 1301[10] (*Tractatus de Ponderibus et Mensuris*)	Same language as above, plus (tower merchant's) pound; units of weight for lead, tallow, and cheese; units for counting quantities of various other trade goods; wool weights: sack of 28 or 30 stones, stone of $12\frac{1}{2}$ pounds	G I Statutes 204–05
33 Edward 1, 1303[11] (*Compositio Ulnarum et Perticarum*)	Barleycorn, inch, foot, yard, perch [rod], acre	G I Statutes 206
9 Edward 3 stat. 1, 1335	First statute mention of "avoir du pois" (meaning "goods of weight")	G I Statutes 269
14 Edward 3 stat. 1 c. 12, 1340	One measure and one weight	G I Statutes 285
14 Edward 3 stat. 1 c. 21, 1340	Weights for wool: sack of 26 stones, stone of 14 pounds	G I Statutes 289

Appendix B. The English statute basis of U.S. weights and measures

Statute	Primary subject of relevant portion	Reference
15 Edward 3 stat. 3 c. 3, 1341	Weights for wool: sack of 26 stones, stone of 14 pounds	G I Statutes 298
25 Edward 3 stat. 5[12] c. 1, 1350	Grain to be sold by striked measure (not heaped)	G I Statutes 319
25 Edward 3 stat. 5[13] c. 9, 1350	Goods to be weighed by balance; weights for wool	G I Statutes 321
25 Edward 3 stat. 5[14] c. 10, 1350	First statute mention of peck, pottle, quart; grain to be sold by striked measure	G I Statutes 321
27 Edward 3 stat. 1 c. 8, 1353	Wines to be gauged; first reference to pipes, tuns, and "the Assize of the Tun"	G I Statutes 331
27 Edward 3 stat. 2 c. 10, 1353	One weight, one measure, and one yard; "avoir du pois" (meaning "goods of weight")	G I Statutes 337
31 Edward 3 stat. 1 c. 2, 1357	Wool weights (sack, half sack, quarter, pound, half pound, quarter [pound]) according to the standard of the Exchequer	G I Statutes 350
31 Edward 3 stat. 1 c. 5, 1357	Wine tuns "to contain a certain number of gallons according to the old gauge"	G I Statutes 350
34 Edward 3 c. 5–6, 1360	Reaffirmations of earlier statutes	G I Statutes 365–66
37 Edward 3 c. 7, 1363	Goldsmiths shall make their silver work of sterling alloy	G I Statutes 380
47 Edward 3 c. 1, 1373	Assise of cloths	G I Statutes 395

Appendix B. The English statute basis of U.S. weights and measures

Statute	Primary subject of relevant portion	Reference
4 Richard 2 c. 1, 1380	Imported vessels of wine, honey, and oil to be gauged	G II 16
5 Richard 2 c. 4, 1381	Prices of wine	G II 18–19
13 Richard 2 stat. 1 c. 9, 1389	Wool weights	G II 63–64
13 Richard 2 stat. 1 c. 10, 1389	Measure of cloths	G II 64
14 Richard 2 c. 11, 1390	Reaffirmation of earlier statutes	G II 77
15 Richard 2 c. 4, 1391	Eight bushels to the quarter (not nine)	G II 79
16 Richard 2 c. 1, 1392	"Avoir de pois" (meaning "goods of weight")	G II 82–83
7 Henry 4 c. 10, 1405	Measure of cloths	G II 154
11 Henry 4 c. 6, 1409	Measure of cloths	G II 163–65
13 Henry 4 c. 4, 1411	Measure of cloths	G II 168
1 Henry 5 c. 10, 1413	Eight bushels striked to the quarter (not nine); the London faat prohibited	G II 174
2 Henry 5 stat. 2 c. 4, 1414	All silver gilt to be English sterling; first statute mention of troy weight	G II 188
9 Henry 5 stat. 1 c. 11, 1421	All gold coin to be received by king's weight	G II 208–09
9 Henry 5 stat. 2 c. 6, 1421	Money to be made according to the same weight used at the Tower	G II 210

Appendix B. The English statute basis of U.S. weights and measures

Statute	Primary subject of relevant portion	Reference
2 Henry 6 c. 14,[15] 1423	Vessels of wine, eels, herrings, and salmon; sizes of the tun, pipe, tertian, hogshead, and barrel	G II 222–23
2 Henry 6 c. 16,[16] 1423	Price of silver; troy pound	G II 223–24
8 Henry 6 c. 5, 1429	Confirmation of 25 Edward 3 stat. 5 c. 9 and 13 Richard 2 stat. 1 c. 9; balance to be used in weighing "toutz maners des choses poisablez"	G II 241–42
9 Henry 6 c. 8, 1430	Wey of cheese; cloves and pounds	G II 267
11 Henry 6 c. 8, 1433	Confirmation of 1 Henry 5 c. 10 etc.; the London faat prohibited	G II 282–84
11 Henry 6 c. 9, 1433	Measure of cloths	G II 284–85
18 Henry 6 c. 16, 1439	Measure of cloths; the "yard and inch"; the "yard and hand"	G II 312–13
18 Henry 6 c. 17, 1439	Vessels of wine, oil, and honey; sizes of the tun, pipe, tertian, and hogshead	G II 313
4 Edward 4. c. 1, 1464	Measure of cloths; the "yard and thumb"; the "yard and nail"	G II 403–07
8 Edward 4 c. 1, 1468	Measure of cloths	G II 424–26
22 Edward 4 c. 2, 1482	Vessels of salmon, herrings, and eels; sizes of the butt, barrel, half barrel, and firkin	G II 470–72
1 Richard 3 c. 8, 1483	Measure of cloths; the "yard and thumb"	G II 484–89
1 Richard 3 c. 13, 1483	Butt of malmsey; sizes of the tun, pipe, tertian, hogshead, barrel, and rundlet	G II 496–97

Appendix B. The English statute basis of U.S. weights and measures

Statute	Primary subject of relevant portion	Reference
7 Henry 7 c. 3, 1491	One measure and one weight	G II 551–52
7 Henry 7 c. 7,[17] 1491	Butt of malmsey	G II 553
11 Henry 7 c. 4, 1494	First attempt to reform standards; quarter, stone, sack, "water measure" (five pecks)	G II 570–72
11 Henry 7 c. 23, 1494	Vessels for salmon and eels	G II 587–88
12 Henry 7 c. 5, 1496	Second attempt to reform standards; definitions of the wine gallon and the troy pound	G II 637–38
6 Henry 8 c. 9, 1514	Measure of wool; first statute mention of "haberdepoys" as a standard of weight	G III 130–31
21 Henry 8 c. 12, 1529	Stone for hemp	G III 291–92
23 Henry 8 c. 4, 1531	Vessels for ale, beer, and soap; barrel, kilderkin, firkin	G III 366–68
24 Henry 8 c. 3, 1532	Beef, pork, mutton, and veal to be sold by "laufull weighte called Haberdepayes"	G III 420
28 Henry 8 c. 14 §5, 1536	Sizes of vessels for wine: tun, butt, pipe, tertian or puncheon, hogshead, tierce, barrel, and rundlet	G III 670
7 Edward 6 c. 7, 1553	Measure of firewood and coal	G III 171
13 Elizabeth 1 c. 11 §4,[18] 1570	Barrel of herrings	G IV 546
23 Elizabeth 1 c. 8 §4, 1581	Barrel, kilderkin, and firkin of honey	G IV 670
31 Elizabeth 1 c. 8, 1589	Imported casks of ale and beer to be gauged by coopers	G IV 805–06
35 Elizabeth 1 c. 6 §8, 1593	Statute definition of the mile	G IV 853
43 Elizabeth 1 c. 14, 1601	Measure of firewood	G IV 981–82

Appendix B. The English statute basis of U.S. weights and measures

Statute	Primary subject of relevant portion	Reference
16 Charles 1 c. 19 §2, 1640	One weight, one measure, one yard; grain to be striked (not heaped)	G V 129
16 Charles 1 c. 19 §6,[19] 1640	Continuation of "water measure"	G V 130
12 Charles 2 c. 23 §8,[20] 1660	Barrel of beer; beer and ale by the ale gallon, other liquors by the wine gallon	G V 256
12 Charles 2 c. 24 §21,[21] 1660	Barrel of beer; beer and ale by the ale gallon, other liquors by the wine gallon	G V 263
14 Charles 2 stat. 2 c. 26 §1, 1662	Weights of kilderkin, firkin, and pot of butter	G V 421
16 & 17 Charles 2 c. 2 §1, 1664-65	Chaldron for coal	G V 552
22 Charles 2 c. 8 §2, 1670	Repeal of "water measure" for grain and salt; Winchester measure only to be used	G V 662
22 & 23 Charles 2 c. 12, 1670-71	Forbidding grain or salt to be sold "by the Bagg"	G V 722
1 William & Mary stat. 1 c. 24 §4,[22] 1688	Barrel of beer or ale	G VI 89
5 William & Mary c. 7 §9,[23] 1693	Cask of pilchards or scads	G VI 449
7 & 8 William 3 c. 31 §43,[24] 1695-96	Salt to be sold by weight rather than by the Winchester bushel	G VII 137
8 & 9 William 3 c. 22 §45, 1696-97	Definition of the Winchester bushel in cubic inches	G VII 256
9 William 3 c. 6 §1,[25] 1697-98	Salt to be sold by weight rather than measure	G VII 302
10 William 3 c. 10 §12,[26] 1698	Barrel of vinegar	G VII 503
11 William 3 c. 15,[27] 1698-99	Measure of ale and beer	G VII 604-05
13 & 14 William 3 c. 5 §9, 1701	"Bushel" means the Winchester bushel	G VII 742

136

Appendix B. The English statute basis of U.S. weights and measures

Statute	Primary subject of relevant portion	Reference
1 Anne stat. 1 c. 9 §1, 1702	"Water measure" for sale of fruit defined; set equal (in effect) to Winchester bushel	G VIII 49
1 Anne stat. 1 c. 15 §6, 1702	Bushel of foreign salt by weight	G VIII 61
1 Anne stat. 1 c. 15 §9, 1702	Bushel of rock salt by weight	G VIII 61
1 Anne stat. 2 c. 3 §6, 1702	Bushel of malt is the Winchester bushel; definition of same	G VIII 156
6 Anne c. 11 art.8, 1706[28]	English barrel for beer and ale in Scots gallons	G VIII 567–68
6 Anne c. 11 art. 17, 1706[29]	Union of Scotland and England; weights and measures to be those of England	G VIII 570
6 Anne c. 27 §22,[30] 1706	Definition of the wine gallon in cubic inches	G VIII 618
8 Anne c. 10 §1, 1709	Chalder for coal	G IX 206
9 Anne c. 6 §5, 1710	Chalder for coal	G IX 367
9 Anne c. 6 §8, 1710	Chalder for cinders	G IX 368
9 Anne c. 17,[31] 1710	Chalder and tun for coal	G IX 473
9 Anne c. 20,[32] 1710	Measure of firewood	G IX 478
10 Anne c. 43, 1711	Chalder for coal	G IX 715
13 Anne c. 20 §13, 1713[33]	Chalder for coal	G IX 970
8 George 2 c. 12 §2, 1735	Local variations in the length of the mile	P XVI 505
24 George 2 c. 31 §8, 1751	Stone for hemp and flax	P XX 224
15 George 3 c. 27, 1775	Bushel and chaldron for coal	P XXXI 45–47

Notes

[1] *1906 Decennial Edition of the American Digest,* 20:842. Here and in the next two citations, the language is that of the *Digest*, not necessarily that of the court.

[2] *1916 Second Decennial Edition of the American Digest,* 23:440.

[3] *1926 Third Decennial Edition of the American Digest,* 27:1476.

[4] Littell, *Statute Law of Kentucky,* 2:572–74; also Littell and Swigert, *Digest of the Statute Law of Kentucky,* 2:1239–41. The same statutes are cited as Title 179 of the laws of Kentucky in 1834 (Morehead and Brown, *Digest of the Statute Laws of Kentucky,* 2:1532–34).

[5] Kilty, *Report of All Such English Statutes as Existed at the Time of the First Emigration* (1811), 289; Alexander, *Collection of the British Statutes in Force in Maryland* (1870), 269 ff.

[6] That is, the *Tractatus de Ponderibus et Mensuris*, traditionally cited as 31 Edward 1 (1301).

[7] Now cited as 2 Henry 6 c. 11.

[8] Now cited as 12 Henry 7 c. 5.

[9] Traditional citation; actual date unknown.

[10] Traditional citation; actual date unknown.

[11] Traditional citation; actual date unknown.

[12] stat. 4 in earlier printed versions.

[13] stat. 4 in earlier printed versions.

[14] stat. 4 in earlier printed versions.

[15] Chapter 11 in earlier printed versions.

[16] Chapter 13 in earlier printed versions.

[17] Chapter 8 in earlier printed versions.

[18] Section 5 in earlier printed versions.

[19] Section 7 in earlier printed versions.

[20] Section 20 in earlier printed versions.

[21] Section 34 in earlier printed versions.

[22] Section 5 in earlier printed versions.

[23] 5 & 6 William & Mary c. 7 §10 in earlier printed versions.

[24] Section 44 in earlier printed versions.

[25] 9 & 10 William 3 c. 6 §1 in earlier printed versions.

[26] 10 & 11 William 3 c. 21 §15 in earlier printed versions.

[27] 11 & 12 William 3 c. 15 in earlier printed versions.

[28] 5 & 6 Anne c. 8 art. 8 in earlier printed versions.

[29] 5 & 6 Anne c. 8 art. 17 in earlier printed versions.

[30] 5 Anne c. 27 §17 in earlier printed versions.

[31] Chapter 22 in earlier printed versions.

[32] Chapter 15 in earlier printed versions.

[33] 12 Anne stat. 2 c. 17 in earlier printed versions.

Glossary of unit names

The following lists all of the Customary units referred to in the text together with the relevant units of the metric system and a few ancient units that play an especially prominent role in the history presented in this book. Etymologies have been compiled primarily from the OED and from the *American Heritage Dictionary* (1st ed.) supplemented in places with *Webster's New International Dictionary* (2d ed.) and various dictionaries of other languages. The descriptions are based partly on the author's opinions and should not be relied upon as definitive; in particular, little attention has been paid to the many other meanings these words have borne over the centuries, and the reader is urged to consult the OED for further information on these uses. **Bold face** type indicates words listed elsewhere in the Glossary. All the words listed here appear also in the Index, which should be consulted for references to fuller discussions. Metric equivalents, if inexact, are given to as many places of precision as allow the figure to be truncated without rounding off (typically at the first zero); these are indicated by a trailing ellipsis. Equivalents without this ellipsis are exact. See the entry for **yard** for an explanation of the definitional bases used to make the calculations.

Abbreviations

cm	Centimeters
cm³	Cubic centimeters
in³	Cubic inches
mL	Milliliters
Gk	Greek
IE	Indo-European
OE	Old English
OF	Old French
L	Latin
ME	Middle English
ML	Medieval Latin
VL	Vulgar Latin

Acre (ME *aker*, OE *æcer* or *acer*; many cognates in the Germanic languages; related to L *ager* "field" and Gk ἀγρός. Ultimately from IE **agro-* "field," derivative of **ag-* "to drive"; original meaning "place where cattle are driven") Fundamental Customary measure of land area, equal to a strip of land one **chain** wide by one **furlong** in length, i.e., 4 **rods** by 40 rods; conceptually, the amount one man with one team can plow in one day. The original meaning of the English word was "a piece of arable land; a field." Since the chain is 66 feet and the furlong is 660, an acre contains 43,560 square **feet** (or 36,000 square Northern feet—see text). In determining the exact metric equivalent of the acre, it must be remembered that the "survey foot" used for land measure is not based on the **inch** of 2.54 cm but rather on the relationship set in 1893 that defined the **yard** in the U.S. as $\frac{3600}{3937}$ **meters**. Thus the acre is 0.4046872609874252... **hectares**.

Acre-breadth (ME *aker brede*) the width of an **acre**—4 **rods**. Later known as a **chain**. ME also had *aker lengthe*, i.e., 40 rods, or a **furlong**.

Are (French, from L *ārea*) The fundamental unit of area in the metric system, equal to 100 square **meters**.

Barrel (OF *baril*, medieval L *barile, barillus, baurilis*; cognates in Provençal, Portuguese, and Spanish. Ultimate origin unknown; perhaps from OF *barre*, late L *barra* "bar," referring to the staves) Cask measure varying in size from 30 to 32 **gallons** in the wine gallon tradition and up to 36 gallons in the ale gallon and Imperial gallon traditions; sometimes defined by weight, as the barrel of flour = 196 **pounds**.

Beqa (From Hebrew inscriptions on some weights) Name applied by Petrie to an Egyptian gold

Glossary of unit names

shekel, considered here to have a normalized value of 192 troy **grains**.

bu/ac **Bushels** per **acre** as a measure of agricultural productivity; that is, bushels by weight produced per acre of area. The bushel in this context is a *weight* that depends on the product as defined by state law, e.g., 60 (avoirdupois) **pounds** per bushel of wheat, 56 pounds per bushel of rye, 48 pounds per bushel of barley, etc. Compare with the bushel as a measure of capacity under **Bushel** and **lb/bu**, and see **kg/ha** for conversion factors.

Bushel (ME *busshel, boyschel, buyschel*, OF *boissiel*; possibly from Gaulish **bostia* "handful," or from OF *boiste* "box," LL *bustellus* "small box," or from OF *boise* "butt") Fundamental Customary measure of capacity for dry substances, equal in the U.S. to the Winchester bushel of 2150.42 in^3, a definition that has remained unchanged for over 300 years. By current definition of the **inch**, the U.S. bushel is equivalent to 35.23907016688 **liters** (or 0.3523907016688 **hectoliters**). The Imperial bushel was defined (1824) in terms of weight of water, originally considered equivalent to 2218.192 in^3, later (1889) 2219.704 in^3; the definition in terms of weight was abandoned in 1985 and the Imperial bushel redefined as 36.36872 cubic **decimeters** (i.e., liters) or 2219.35546233297191... in^3. Compare the definition of the Winchester bushel as a certain volume with the bushel as the weight of a particular agricultural product under **bu/ac**. The Winchester bushel remained in use in Canada for almost a century after it had been replaced by the Imperial bushel in Britain.

Butt (ME *butte*, OF *boute* "butt," late L *butta, buttis* "cask," from the diminutive of which, *butticula*, we get *bottle*; of unknown origin) Cask measure of 126 wine **gallons**; same as the **Pipe**.

Centiliter A hundredth of a **liter**; 10 cm^3.

Centimeter A hundredth of a **meter**.

Chain Surveyor's name for the **acre-breadth** (4 **rods** or 66 **feet**); a tenth of a **furlong**. Introduced by Edmund Gunter ca. 1620.

Chalder Shortened form of **chaldron** (see next), used in England as the name of a measure for lime and coal.

Chaldron (Variant of *cauldron*. OF *chauderon*, from *chaudère, chaudière* "kettle"; L *caldāria* "cauldron," plural of *caldārium* "hot-bath," from *caldus, calidus* "hot"; IE **kel-* "warm") In Winchester measure, 4 **quarters** or 32 **bushels**.

Clove (L *clāvus* "nail," IE **kleu-* "hook, peg"; frequent in laws and ordinances of the 13th–15th centuries in the Anglo-Latin form *clavus* and the Anglo-French form *clou*, but it is not clear how the word for "key" became the name of a weight. See **nail** below) A weight for wool and cheese; in the avoirdupois system of weight for wool, 7 **pounds**.

Cubit (ME and OF *cubite*, L *cubitum*, "elbow, distance from elbow to finger-tips," related to L *cubāre* "to lie down, recline"; IE **keu-* "to bend"; cognates in a number of Romance languages) Generic archaeological name for any of various ancient units of length corresponding roughly to the length of the forearm, sometimes (as in Latin, Greek, and Hebrew) called by a name meaning "forearm" and sometimes not, and sometimes actually close to the length of an adult human forearm and sometimes quite a bit longer than that. For details on the common Hebrew cubit of 17.68 inches, the Egyptian royal cubit of 20.62–20.63 inches, and the Sumerian cubit of 19.49 inches, see the Index. The Roman foot of 11.64–11.65 inches and the Greek foot of 12.44 inches both had cubits $1\frac{1}{2}$ times the length of the corresponding foot. This was the most common relationship between the cubit and the foot in ancient systems (foot = $\frac{2}{3}$ cubit), but we also find foot = $\frac{3}{5}$ cubit and foot = $\frac{1}{2}$ cubit (as for example the Northern cubit mentioned in the text). Other ancient cubit standards (see SKINNER) include the the Olympic cubit of 18.23 inches, the Black cubit of 21.28 inches, the Assyrian cubit of 21.4–21.6 inches, the Persian cubit of 25.2–25.3 inches, and the Hashimi cubit of 25.56 inches. English dictionaries often cite an English cubit of 18 inches, but there appears to be no evidence of this cubit actually being used as a unit of measure in the Customary tradition; the word exists in English only to translate the name of the unit from other languages.

cwt Abbreviation of **hundredweight** (Latin *centum*, "hundred").

Deben Ancient Egyptian unit of weight equal to 10 **qedets**, considered here to have a normalized value of 1440 troy **grains**.

Decimeter One tenth of a **meter**; 10 cm.

Denarius Roman silver coin whose name became adopted in a number of later cultures to mean the most common silver coin (for example, Arabic *dinar*) or money in general (for example, Spanish *dinero*). The word was commonly used in old English statutes written in Latin to mean the silver penny, hence the abbreviations *d* for pence and **dwt** for **pennyweight**.

Didrachma (Double **drachma**) Name of various ancient coins whose weight was based on a **shekel**.

Digit (L *digitus* "finger"; IE **deik-* "to show" in the sense of pointing) The width of a **finger**. The Greek foot was divided into 16 digits and the Roman foot into both 16 digits and 12 **inches**, as was our Customary foot in times past (digit or finger = $\frac{3}{4}$ inch or 1.905 cm). The word *digit* is also commonly used by archaeologists to refer to roughly similar traditional units that probably were not thought of as fingerbreadths; for example, the Sumerian measure known in English as a foot (at 33.0 cm even less representative of a real foot than ours is) was divided into 20 "digits" that were actually rather narrower (1.65 cm or 0.65 inch) than other digits properly so called.

Drachma (L *drachma*, Gk δραχμή "drachma," earlier "handful"; IE **dergh-* "to grasp") The principal silver coin of the ancient Greeks; generically, the weight of half a **shekel**. The name (but not the weight) was later applied to the Arabic *dirhem* and the Customary **dram**. The drachma of most interest in this book was an Attic Greek weight unit of $67\frac{1}{2}$ troy **grains**, $\frac{1}{100}$ of the Attic **mina** that became the medieval English *libra mercatoria* or tower merchant's **pound**.

Dram (British *drachm*, Medieval French *drame*, *drachme* "dram, eighth part of an ounce"; from **drachma**) A Customary unit of weight and capacity. The troy dram (apothecaries' dram) is $\frac{1}{8}$ troy **ounce** or 60 troy **grains** (3.88793460 **grams**); the avoirdupois dram is $\frac{1}{16}$ avoirdupois ounce or $27\frac{11}{32}$ troy grains (1.7718451953125 grams). The fluid dram is $\frac{1}{8}$ fluid ounce (0.2255859375 in^3 or 3.6966911953125 cm^3).

dwt Abbreviation for **pennyweight**. See **Denarius**.

Ell (ME *eln*, OE *eln* "cubit", Old Teutonic **alina* "forearm," with descendants in a number of European languages, including French *aune*; cognate with L *ulna*, Gk ὠλένη, having the same meaning; all from IE **el-* "forearm"—the *el* in *elbow*) Semantically, the ell is identical to the **cubit**, but metrologically the English ell is actually a double cubit, just as the **yard** is. Indeed, the word (as *ulna*, *eln*, or *aune*) was used in Anglo-Latin and Norman French to refer to the yard. Later ell became the name of a Customary cloth measure, no longer used, of 45 inches. The cloth ell is 5 **spans** or 20 **nails**, whereas the yard is 4 spans or 16 nails.

Faat (A variant of *vat*: OE *fæt*, Germanic **fatam* "vessel," with cognates in a number of Germanic languages; IE **ped-* "container") Also spelled *fat*, *fatt*, and *vaat*. An old measure of 9 **bushels**, representing a **quarter** (8 bushels) heaped up rather than striked. The unit appears in three old statutes attempting (apparently unsuccessfully) to prohibit its use, beginning as a ban against "nine bushels per quarter" (*noef busselx p^r le quartre*) in 15 Richard 2 c. 4 (1391) and then by name in 1 Henry 5 c. 10 (1413) and 11 Henry 6 c. 8 (1433). The struggle between the legal desire to accurately define capacities by striked measure and the psychological desire to buy goods heaped up (well known to the marketers of fast foods) is a constant note in the history of English standards.

Glossary of unit names

The metrological significance of the heaping ratio of $\frac{9}{8}$ (which explains a number of details in the history of weights and measures, such as the commonly seen ratio of $\frac{3}{2}$ between the **cubit** and the **foot**) will be explored in a subsequent study.

Fathom (ME *fadme*, OE *fæðm*, Germanic **fathmaz* "length of two arms outstretched," IE **pet-* "to spread") The distance from fingertip to fingertip of the outstretched arms; Customary unit of length for soundings and other nautical uses, 6 **feet**.

Ferlingate (ME *ferling*, OE *fēorthling* "a fourth part") Anglo-Saxon name for the square **furlong**. The generally accepted derivation of *ferlingate* from *ferling* "a fourth" is explained by the fact that a square furlong (10 acres) is one-fourth of a **virgate** of 40 acres; but it's equally true that the virgate was sometimes 30 acres or some other area depending on the quality of the land, and *ferling* looks enough like *furlong* to suggest that a word meaning "square furlong" might have an etymology different from the one given in the dictionaries.

Fifth In the U.S., a fifth of a gallon; formerly the standard size of a bottle of wine or spirits, now generally replaced by the bottle of 750 **milliliters**.

Finger (OE *finger*; Germanic **fingwraz* "one of five," IE **penkwe* "five") Customary version of the **digit**; $\frac{3}{4}$ **inch** (1.905 cm).

Firkin (Earlier *ferdekyn*, apparently from Middle Dutch **vierdekijn*, diminutive of *vierde* "fourth, fourth part") Cask measure equal to a fourth of a **barrel** or half a **kilderkin**.

Foot/feet (ME *fot/fet*, OE *fōt/fēt*, Germanic **fōt-*, IE **ped-* "foot") Generic English word for length units in the range of 10–14 **inches**, whether referred to by the native word for "foot" or not. In the Customary system, a unit of 12 inches, the third of a **yard**. In the U.S., the foot has been defined since 1893 as $\frac{1200}{3937}$ meters. Since 1959, a unit called the *international foot* has been defined as 0.3048 meters; the name is something of a joke, as the United States is the only country that uses it. In practice, the international foot is useful only in giving us a simplified definition of the inch as 2.54 cm. See **yard** for a description of the convention adopted here for calculating metric equivalents.

Furlong (OE *furlang*, from *furh* "furrow" (Germanic **furh-*, IE **perk-* "tear out, dig out") + *lang* "long" (Germanic **langaz*, IE **del-* "long"); originally, the length of the furrow in the common field, a square containing 10 **acres**) A basic Customary unit of land measure, equal to 40 **rods** or $\frac{1}{8}$ **mile**. In the U.S., the furlong is $\frac{792000}{3937}$ or $201.1684023368\ldots$ **meters**.

Gallon (Old North French *galun*, *galon*, Central OF *jalon*, ML *galōnem*; apparently cognate with French *jale* "bowl," OF *galaie*, *galeie*, *jalaie* "measure for liquids, grain, etc.," ML *gallēta* "jug, measure for wine"; ultimate origin unknown, possibly from Celtic) Fundamental Customary unit of liquid capacity, equal to 231 cubic **inches** (3.785411784 **liters**). There were four gallon standards in the English tradition: the Guildhall gallon of about 224 in^3, the wine gallon of 231 in^3, the Winchester gallon of 268.8025 in^3 (4.40488377086 liters), and the ale gallon of about 282 in^3. The wine gallon and Winchester gallon became our U.S. Customary units of liquid and dry capacity, respectively. The British Imperial gallon, defined in 1824, was later adopted in Canada but has never been used in the U.S.; for metric equivalents of Imperial measure, see **Bushel** (gallon = $\frac{1}{8}$ bushel).

Gill (OF *gille*, *gelle*, ML *gillo*, *gellus*, a vessel or measure for wine. Apparently related somehow to *gallon*, though the derivation is not clear) Customary unit of liquid measure, $\frac{1}{4}$ **pint** or 4 fluid **ounces**. Pronounced "jill" and in older documents sometimes spelled that way.

Grain (OF *grain*, *grein*, L *grānum* "grain, seed," IE **grə-no-* "grain") The basic small unit in virtually all traditional systems of weight. Prehistorically the weight of the grain must have depended on the species of plant from which the seed came (as it did up till the 19th century in places like India), but by historical times, the weight of the grain in most places had become a specified subdivision (for example, $\frac{1}{180}$) of a standard **shekel**. In England, the grain was by ancient tradition considered to be a grain of wheat weighing $\frac{1}{640}$ **ounce**, but the grain used in practice seems always to have been the troy grain, notionally a grain of barley and then, as now, $\frac{1}{480}$ of a troy ounce. The troy grain is equal to 64.79891 **milligrams** (exactly).

Gram (French *gramme*, late Latin *gramma* "small unit," Gr γράμμα "written letter," IE *gerebh- "to scratch") Metric unit of mass/weight, $\frac{1}{1000}$ of a kilogram. A gram weighs 15.43235835294143... troy grains.

Hand (OE *hand*, Germanic *handuz*) Customary unit of length equal to 4 **inches**, traditionally used to specify the height of horses.

Hectare One hundred **ares** or 10,000 square **meters**. A hectare is 2.47104393046627895... **acres**.

Hectoliter One hundred **liters**; the metric unit for grains and other substances that in the Customary system would be measured by the **bushel**. 1 hectoliter (hl) is 2.83775932584... U.S. bushels.

Hide (OE *hīgid*, *hīd*; apparently from *hīwid*, a derivative of *hīw-* "household, family," IE *kei- "home"; original meaning possibly "homestead") Anglo-Saxon measure of land area, conceptually the amount that could be tilled with one plow in one year, or the amount sufficient to support one free family and its dependents. The actual acreage varied according to the nature of the land, most commonly either 3 or 4 **virgates**, i.e., 120 or 160 **acres**, the latter being the figure used in the Domesday Book.

Hogshead (Just what it looks like—the head of a hog. The trope was picked up in Danish *oxehoved*, which means "ox head," and similar words in Flemish, Dutch, Swedish, and German, but no one seems to know why) Customary cask measure equal to 2 **barrels**.

Hundredweight In the U.S., 100 pounds, but historically 112 pounds (in the traditional avoirdupois series of weights for wool) and in medieval English contexts often 120 pounds. Abbreviation **cwt**.

Inch (OE *ynce*, L *uncia* "inch, ounce, twelfth part (of a pound or a foot)" from *ūnus* "one", IE *oino- "one." The word *inch* represents an early adoption into English from Latin and is not found in most other European languages, which typically use a word meaning "**thumb**") One twelfth of a foot. The U.S. inch is now defined as 2.54 cm (exactly), and this is the definition used in this book to calculate all the metric equivalents to capacities in cubic inches (1 in^3 = 16.387064 cm^3).

Kilderkin (ME *kilderkyn*, Middle Dutch *kinderkin*, *kinnekijn*, diminutive of *kintal* "hundredweight," from ML *quintāle*, from Arabic *quintāre*, from Late Latin *centēnārium* (*pondus*) "hundredweight", *centum* "hundred", IE *dkm-tom-* "hundred") Customary cask measure equal to half a **barrel** or 2 **firkins**.

kg/ha Kilograms per **hectare**; the metric equivalent of the Customary **bu/ac** (bushels per acre) as a measure of agricultural productivity. In converting between bu/ac and kg/ha, it's important to note that kg/ha is absolute regardless of the agricultural product in question, whereas bu/ac, while also a weight of product per area of land, varies depending on the product because of the different definitions of the bushel as a weight (see bu/ac). Thus, 1 bushel of wheat per acre (bushel of wheat = 60 avoirdupois pounds) is 67.25080036765888... kilograms of wheat per hectare, whereas 1 bushel of barley per acre (bushel of barley = 48 avoirdupois pounds) is $\frac{4}{5}$ of this or 53.8006402941271... kilograms of barley per hectare.

kg/hl Kilograms per **hectoliter**; the metric equivalent of the Customary **lb/bu**. To convert a quantity expressed in kg/hl to its equivalent in lb/bu, multiply by 0.7768885126281996...; to convert a quantity expressed in lb/bu to kg/hl, multiply by 1.287185978097447...

Glossary of unit names

Kilogram Fundamental unit of weight/mass in the metric system, originally the weight of a cubic **decimeter** of water *in vacuo* at its temperature of maximum density; since 1889 defined simply as the mass of the kilogram standard (the International Prototype Kilogram) housed at the International Bureau of Weights and Measures in Sèvres, France. A kilogram is 2.2046226218487758... **pounds** avoirdupois.

Kilometer One thousand **meters**. A kilometer is $\frac{3937}{6336}$ or $0.62136994\overline{49}$ U.S. miles.

lb/bu Pounds per **bushel**; that is (in the U.S.), avoirdupois pounds of 7000 troy **grains** per Winchester bushel of 2150.42 in^3.

Link One hundredth of a surveyor's **chain**, 7.92 **inches**. An **acre** is 100,000 square links. The measure is now obsolete.

Liter (French *litre*, from *litron*, an old French measure of capacity; apparently from late Latin *litra*, from Gk λίτρα "pound"; Mediterranean *līthra* "scale") Fundamental metric unit of capacity, originally defined as the volume of a cubic decimeter, then (1901) as the volume of a kilogram of water at standard atmospheric pressure and the temperature of its maximum density (about 4 °C); this would make the liter equal to about 1.000028 cubic decimeters. In 1964 this definition was abolished and the liter defined again as another name for the cubic decimeter. The result of all this confusion has been the official abandonment of the liter as a unit of measure for scientific purposes. According to its current definition, 1 liter is equal to 61.02374409473228395... in^3, which is 1.05668820943259366... U.S. liquid quarts or 0.90808298426885589... U.S. dry quarts.

Mark or marc (OE *marc*, ML *marca, marcus*; the word exists in all of the Germanic and Romance languages, and authorities differ as to whether it comes from some Romanic root or from Germanic **mark-* "boundary, mark" (in this connection, presumably, a mark upon a bar of metal), from IE **merg-* "boundary, border") Generic traditional European unit of weight for gold and silver, equal to 8 **ounces** and thus to $\frac{2}{3}$ **pound** or $\frac{1}{2}$ pound depending on whether the pound contains 12 or 16 ounces. The word can still be seen in *Deutschemark*, the name of the German unit of currency until its recent replacement by the euro.

Meter (French *mètre*, from Gk μέτρον "measure") Fundamental unit of length in the metric system, intended originally to equal one ten-millionth of the length of a quadrant of the earth's meridian but not achieving it. It is currently defined in terms of the wavelength of a spectral line of an isotope of krypton. Land measure in the U.S. is based on the relationship 1 meter = $\frac{3937}{3600}$ (1.0936$\overline{1}$) **yards**, whereas scientific measurements are based on the relationship 1 foot = 0.3048 meters; the difference between the two definitions is two parts in a million.

Mile (OE *mīl*, West Germanic **mīlja*, L *mīllia, millia*, plural of *mīle, mille* "thousand" (perhaps from IE **ghesio-* "thousand") in the Latin term *mille passus* or *passuum* "a thousand paces," the **pace** being 5 **feet**. The word in some form is common to many European languages) Customary unit of itinerary measure, equal to 8 **furlongs** or 320 **rods**. In the U.S., a mile is equal to $\frac{6336}{3937}$ or 1.6093472186944... **kilometers**. There is also a variant called the *international mile* that is equal to 1.609344 kilometers (exactly); it has been superseded by the kilometer for all practical uses and is best forgotten.

Milligram A thousandth of a **gram**. A milligram is equal to 0.01543235835294143... troy **grains**.

Milliliter A thousandth of a liter. Originally a cubic **centimeter**, then the volume of a **gram** of water, and now back to being a cubic centimeter again; see **liter**.

Millimeter A thousandth of a **meter**.

Mina (L *mina*, from Gk. μνα, Hebrew *maneh*; from Akkadian *manū*, from Sumerian *mana*) Generic name for ancient units of weight equal to about a **pound**. It was common in ancient cultures also to have a "king's mina" or "royal mina" that was exactly double the weight of the mina of commerce. The mina was typically divided into 25 to 60 smaller units (**shekels**) depending on the system. See also **talent**.

Minim (L *minimus* "smallest", IE **mei-* "small") The smallest Customary unit of liquid measure, about the size of one drop; $\frac{1}{60}$ fluid **dram** or $\frac{1}{480}$ fluid **ounce**. In the U.S., the minim is 0.003759765625 in³ or 0.061611519921875 cm³.

Nail (ME *nail*, OE *nægl*, IE **nogh-* "nail, claw") A Customary measure of length, $\frac{1}{16}$ **yard** or $2\frac{1}{4}$ **inches**. Markings for the nail can still be found along the yard measures set into the counters of fabric stores. The OED confesses itself unsure of the association between the fastener called a nail and the length called a nail, but it's reasonable to guess that it comes from the practice of marking the divisions of the yard with small nails set into the shop counter. The cloth nail is $\frac{1}{4}$ of a **span** and $\frac{1}{20}$ of an **ell**. *Nail* is also another name for the obsolete weight known as the **clove** (7 pounds for wool and 8 pounds for cheese); the rationale for the use of the word to refer to a weight is even more obscure than the rationale for its use to refer to a length.

Ounce (ME *unce*, OF *unce*; same derivation as **inch**) Customary units of weight and capacity. As a unit of weight, either the troy ounce ($\frac{1}{12}$ troy **pound**) or the avoirdupois ounce ($\frac{1}{16}$ avoirdupois pound); both are discussed at length in the text. The troy ounce of 480 troy **grains** is equal to 31.1034768 **grams**, and the avoirdupois ounce of $437\frac{1}{2}$ troy grains is equal to 28.349523125 grams. As a unit of capacity (the fluid ounce), $\frac{1}{128}$ U.S. gallon, equal to 1.8046875 in³ or 29.5735295625 cm³. The Imperial fluid ounce is $\frac{1}{160}$ of an Imperial gallon or 28.4130625 cm³ (1.7338714549476343... in³).

Pace (ME *pace*, OF *pas*, L *passum, passus*, "pace," literally "a stretch" (of the leg), from *pandĕre* "to stretch, extend"; IE **pet-* "to spread") Customary unit of length, in full, *geometric pace*, equal to 5 **feet**, i.e., two steps, the latter being called the *military pace* ($2\frac{1}{2}$ feet); the Roman *passus* and *gradus* represented exactly the same distinction and were defined the same way with regard to the Roman foot.

Palm (ME *paume*, OF *paume*, L *palma* "palm of the hand"; IE **pele-* "flat; to spread") In our tradition, a rough measure of length taken either from the width of the hand, about 3–4 **inches**, or the whole length of the hand from the wrist to the tip of the longest finger, 7–9 inches. In many traditional European systems of weights and measures, however, the palm (*palma, palmo*, etc.) was a specific unit of length, typically $\frac{1}{4}$, $\frac{1}{3}$, $\frac{2}{3}$, or $\frac{3}{4}$ of the **foot**, or sometimes $\frac{1}{2}$ of the local equivalent of the **cubit**, and the longer version was in some cases a more important unit than the foot. This more precise role of the palm (as width) is played in our system by the **hand** and of the palm (as length) by the **span**.

Peck (ME *pek*, OF *pek*, "fourth of a bushel"; earlier history unknown (possibly from the verb *pekken* "to peck or snap"), but probably related to the root seen in French *picotin*, "a fourth of a bushel of oats to be fed to horses," from ML *picotus, picota*, a liquid measure, OF *picota*, a wine measure) Customary unit for the sale of vegetables (heaped) and grains; a fourth of a **bushel**, or two **gallons**. In the U.S., 537.605 in³.

Penny (ME *peny*, OE *penig, penning*, West Germanic **panninga*, the Germanic derivation evidenced by the appearance of the word in most of the Germanic languages (for example, German *pfennig*); possibly from L *pannus* "a cloth" on the theory that pieces of cloth were used as a medium of exchange in barbarian Europe, from IE **pan-* "fabric"; or possibly from a base **pand* (German *pfand*) "to pledge" + *ing*, i.e., "little pledge, token, coin") The basic medieval English silver coin, worth $\frac{1}{240}$ **pound** (money) and weighing originally $\frac{1}{240}$ of the pound sterling or tower pound of 5400 troy **grains**, i.e.,

Glossary of unit names

$22\frac{1}{2}$ grains. Abbreviation *d* (see **Denarius**).

Pennyweight Unit of troy weight, $\frac{1}{240}$ troy pound of 5760 troy grains, i.e., 24 grains; $\frac{1}{20}$ troy **ounce**. Abbreviation **dwt**.

Perch (ME *perche*, OF *perche*, L *pertica* "pole, measuring rod," Italic **pert-* "pole") Customary unit of area, equal to a square **rod** or $\frac{1}{160}$ **acre**; but see usage note under **Rod**. The perch is $30\frac{1}{4}$ square English **yards** or 225 square Northern **feet** (see Index); this is equal to $\frac{392040000}{15499969}$ or 25.292953811714... square **meters**.

Petroleum barrel International unit of measure for crude oil, equal to 42 U.S. gallons; see **Tierce**.

Pint (ME *pynte*, OF *pinte*, a liquid measure; the word also appears in Italian, Spanish, Portuguese, Dutch and German. Probably from ML *pincta* "painted mark (on the inside of a container)," VL **pinctus* "painted," L *pingere* "to paint," IE **peig-* "to cut, mark") Customary unit of liquid and dry capacity, $\frac{1}{8}$ **gallon** or $\frac{1}{2}$ **quart**. The U.S. liquid pint is 28.875 in³ (473.176473 cm³ or 0.473176473 liters); the U.S. dry pint is 33.6003125 in³ (550.6104713575 cm³ or 0.5506104713575 **liters**).

Pipe (OF *pipe*, a measure for wine, etc. The word originally referred to a musical pipe (L *pipāre* "to chirp") and then by extension to other tubular objects) Cask measure of 126 wine **gallons**; 2 **hogsheads**, 4 **barrels**, or $\frac{1}{2}$ **tun**. Also called a **butt**. Theoretically still a unit of measure in the U.S., though not actually used for a century or more.

Pottle (ME *potel*, OF *potel*, "a little pot"; *pot* (which appears in a number of European languages) is from VL **pottus*, from Celtic **pott-* "pot") Old name for the half **gallon**, or 2 **quarts**. The unit has recently been reborn as the *growler* for beer.

Pound (OE *pund*, from West Germanic stem **pundo-* "pound weight"; very early adoption from L *pondo* "pound weight," short for *libra pondo* "a pound by weight"; *libra* (the Roman pound) + **pondus* "weight," from *pendere* "to weigh"; IE **spen-* "to draw, stretch, spin.") The word *pound* and its cognates (e.g., German *pfund*) are found in most of the Germanic languages) Generic name for descendants of various bronze age **mina** standards, including two Customary units of weight, the troy pound of 5760 troy **grains** (12 troy **ounces** of 480 grains) and the avoirdupois pound of 7000 troy grains (16 avoirdupois ounces of $437\frac{1}{2}$ grains); see the text for a fuller discussion. The troy pound is equal to 373.2417216 **grams** and the avoirdupois pound to 453.59237 grams.

Puncheon (OF *ponçon, ponchon*; origin unknown. *Puncheon* meaning "post" or "tool for piercing" (OF *poinchon*, modern French *poinçon*) is related to *punch* and is considered by lexicographers of English and French to be a separate word; but Italian lexicographers consider them the same word, the idea being that a punch was used to mark or seal a cask) Old Customary cask measure of 84 wine **gallons**; $\frac{1}{3}$ **tun** or 2 **tierces**. Theoretically still a unit of capacity in the U.S., though not used since the 19th century. Also called a **tertian**.

Quart (OF *quarte*, "fourth part (of a gallon)," L *quārtus*, "fourth"; IE **kwetwer-* "four") One fourth of a **gallon** (in all contexts); 2 **pints**.

Quarter (OF *quarter, quartier*; see **quart**) As capacity, a measure of grain equal to 8 **bushels**, or a quarter of a **chaldron**; as weight, a quarter of a **hundredweight** (i.e., either 25 or 28 **pounds**).

Qedet (New Kingdom Egyptian *qdt*, vocalized for scholarly purposes as *qedet*) The ancient Egyptian **shekel** of commerce, estimated most accurately from the shipwreck at Uluburun to weigh 9.32–9.35 **grams** (143.8–144.3 troy **grains**) and considered in this book to be the ancient basis of the troy standard of weight at a normalized value of 144 grains (9.33 grams), i.e., $\frac{1}{40}$ troy **pound** or 6 **pennyweights**. See **deben** and **sep**.

Rod (ME *rodd*, OE *rodd*; IE **rēt-* "post") Fundamental Customary unit of length for land measure, originally 15 Northern **feet** (see Index) and redefined in the 13th century (without changing its length) as $5\frac{1}{2}$ English **yards** or $16\frac{1}{2}$ English feet. In the U.S., the rod is equal to $\frac{19800}{3937}$ or 5.02921005842... **meters**. *Usage note:* In this book, *rod* always means the unit of length and *perch* always means the corresponding unit of area (a square rod, see **perch**); but historically, the distinction is slight, the word *perch* in old books and manuscripts often being used to refer to the length and the word *rod* (a little less frequently) to refer to the area. The word *pole* is also sometimes used with these meanings. The convention adopted here reflects the general use of *rod* to refer to the unit of length in U.S. state laws and land surveys.

Rood (ME *rood*, OE *rōd*, "rod, cross"; basically the same word as **rod**) Old Customary unit of land area, equal to $\frac{1}{4}$ **acre** or 40 **perches**.

Rundlet (From OF *rondelet* "little cask," diminutive of *rondelle*, from *ronde* "round," from VL **retundus*, L *rotundus* "round," related to *rota* "wheel"; IE **ret-* "to run, roll") Old cask measure, defined by statute as 18 or $18\frac{1}{2}$ wine **gallons**.

Sack (ME *sack, sak*, OE *sæcc, sacc*, from L *saccus* "bag, sack," from Gk σάκκος, adapted from Semitic *saq* (Jewish Aramaic *saq, saqqā*, Syriac *saq, saqa*, Assyrian *saqqu*), "sack, sackcloth"; the word is found in most European languages as a borrowing from either L or Gk) A unit of measure for grain, wool, etc.; specifically, a weight of wool equal to 26 **stone** or 364 avoirdupois **pounds**.

Sarpler or sarplier (OF *sarpilliere*; origin unknown) A sack or bale of wool weighing originally 2 **sacks** (728 avoirdupois **pounds**); later, according to the OED, equivalent to 80 **tods** (160 **stone**) or 2240 avoirdupois pounds, i.e., 1 **ton**.

Scruple (OF *scrupule*, from L *scrūpulum, scrĭpulum*, etc. "small weight, small stone") A unit of apothecaries' weight, 20 troy **grains** ($\frac{1}{3}$ **dram** or $\frac{1}{24}$ troy **ounce**).

Sep Ancient Egyptian unit of weight, equal to 10 **debens** or 100 **qedets**.

Shaftment (OE *sceaftmund*, from OE *sceaft* "shaft" (Germanic **skaftaz*, IE **skep-* base of words associated with cutting and scraping) + OE *mund* "hand, handbreadth," IE **man-* "hand") "The distance from the end of the extended thumb to the opposite side of the hand" (OED). The image is that of a hand grasping a rope or pole with the thumb extended along the object grasped; then the length in shaftments is counted by working along hand over hand.

Shekel (Hebrew *sheqel*, from *shāqal* "to weigh"; Akkadian *shiqlu* "weight") Generic name for any of a number of weight standards of the ancient Near East, the most important being the Palastinian/Hebrew peyem of 120 troy **grains**, the Sumerian shekel of 129.6 grains, the Attic **stater** of 135 grains, the Egyptian **qedet** of 144 grains, and the Egyptian **beqa** of 192 grains. See **mina** and **talent**.

Shilling (ME *shilling*, OE *scilling*, from Germanic **skillingaz*; common in most European languages) A coin or weight equal to $\frac{1}{20}$ pound (money or weight). As a coin, equal in value to 12 **pennies**; as a weight, equal in weight to 12 **pennyweights**.

SI Système International (d'Unités) or International System of Units; the official name of the metric system.

Span (ME *span(ne)*, OE *span(n), spon(n)*, apparently from the verb *spannan* "to fix, fasten, join" from Middle Dutch *spannen* "to join," Germanic **spannjan*, IE **spen-* "to draw, stretch, spin") The distance from the tip of the thumb to the tip of the little finger; Customary unit of length equal to 9 inches, i.e., $\frac{1}{4}$ **yard** or $\frac{1}{5}$ **ell**.

Stater (L *stater*, from Gk στατήρ, from στα-, ἱστάναι "to place in the balance, weigh," IE **stā-* "to stand") Ancient Greek weights, in particular the Argive standard equal to 2 troy **pounds**; also, various ancient coins defined as fractions of the larger weights, in particular the Attic silver stater of 135 troy **grains**.

Stone (ME *stane, stone*, OE *stān*, Germanic **stainaz*, IE **stei-* "stone") Unit of weight, $\frac{1}{8}$

Glossary of unit names

hundredweight or ½ **quarter**; in avoirdupois weight, 14 **pounds** (the hundredweight being 112). *Stone* as a unit of weight is both singular and plural.

Tablespoon (18th century) Customary cooking measure, equal to 4 liquid **drams** or ½ fluid **ounce** (0.90234375 in^3, 14.78676478125 cm^3 or mL). Often incorrectly but harmlessly equated with the metric tablespoon (15 mL). The word for the measure of capacity should properly be *tablespoonful,* but this has virtually disappeared in current usage.

Talent (OE *talente, talentan,* L *talenta* (plural of *talentum*), adapted from Gr τάλαντον, "balance, weight, sum of money"; related to L *tollere* "to lift." The use of the word to mean "mental endowment, natural ability" comes from the parable of the talents, Matthew 25:14–30) Generic name for the basic ancient unit of weight, equivalent to the load that a person can carry short distances (about 60 **pounds**). The standard talents referred to in the text (see Index) were the Aeginetan talent, considered here to have a normalized weight of 576,000 troy **grains**; the Syrian talent, considered to have a weight of 432,000 grains; and the Attic talent, considered to have a weight of 405,000 grains. Talents were generally divided into 60 **minas**, but the further division of the mina into 25 to 60 **shekels** depended on the weight system. Most commonly the talent consisted of 3600, 3000, or 2400 shekels, representing divisions of the mina by 60, 50, and 40, respectively.

Teaspoon (18th century) Customary cooking measure, equal now to ⅓ **tablespoon** (⅙ fluid **ounce**) but until the 1930s the same as the fluid **dram** (¼ tablespoon). The teaspoon thus contains 0.30078125 in^3 (4.92892159375 cm^3 or mL). The Customary teaspoon is often incorrectly but harmlessly equated with the metric teaspoon of 5 mL. The word for the measure of capacity should properly be *teaspoonful,* but this has virtually disappeared in current usage.

Tertian (L *tertius* "third"; L *trēs,* IE **trei-* "three") A third of a **tun**, 84 wine **gallons**; same as the **puncheon**.

Tetradrachma (Quadruple **drachma**) An ancient Greek silver coin weighing 4 **drachmas**.

Thumb (ME *thom(b)e,* OE *þūma,* Germanic **þūmon-,* pre-Germanic **tumon-* "the thick (finger)," IE **teue-* "to swell") The word used in most Germanic languages for the unit we call the **inch** (e.g., Danish *tomme,* Dutch *duim;* but oddly, not German, which has *zoll,* from Middle High German *zol,* "cylindrical piece, log"). Romance languages use the corresponding word for "thumb" from Latin, e.g., Spanish *pulgada* "inch" from VL **pollicata,* L *pollicaris* "size of a thumb," *pollex, pollicis* "thumb"; Italian *pollice,* French *pouce.* English appears to be rare or possibly unique in anciently having adopted the actual Latin word for "inch" (*uncia*) as the name for this unit.

Tierce (OF *terce, tierce,* L *tertium, tertiam* "third"; see tertian) A third of a **pipe**, i.e., 42 wine **gallons**. Adopted by the early U.S. petroleum industry and recognized internationally as the **petroleum barrel**.

Tod (ME *todd(e),* probably from a Germanic word **toddōn* meaning something like "tuft or bunch," later "bundle, small load") Wool weight equal to 2 **stone**.

Ton (From *tun,* and until the late 17th century spelled that way) A unit of weight equal to 20 **hundredweights**; thus, either 2240 avoirdupois **pounds** (as in Britain) or 2000 pounds (as in the U.S.) depending on the definition of the hundredweight. When distinguishing between the two ton weights, the larger is called the *long ton* and the smaller is called the *short ton*. The word *ton* is also used to refer to capacities for lumber and some other commodities; in shipping, the volume of 100 cubic feet (for registered tonnage) or 40 cubic feet (for freight). The French spelling of the same word, *tonne,* is sometimes used in English to refer to the metric ton, 1000 **kilograms**; the metric ton is equal to 2204.6226218487758... pounds, i.e., 1.1023113109243879... short tons or 0.98420652761106... long tons, and the tons of 2000 and 2240 pounds are equal to 0.90718474 and 1.0160469088 metric tons, respectively.

Tun (ME *tunne,* OE *tunne;* found in most European languages, both Germanic and Romance, including ML (*tunna*), but apparently not originally from Latin but rather from Celtic, the word being akin to Middle Irish *tunna,* Old Irish *toun* "hide, skin," which would make the original meaning something like "wine skin") The largest of the Customary cask measures, equal to 2 **pipes**, 4 **hogsheads**, or 8 **barrels**; 252 wine **gallons**.

Virgate (ML *terra virgāta,* used as a rendering of OE *gierd-land* "**yard**-land" and having the same meaning; from L *virga* "rod") An old unit of land area, one fourth of a **hide**; nominally 40 **acres**, but varying in practice depending on the quality of the land.

Wine glass (19th century) More properly *wine-glassful:* 2 fluid **ounces**.

Yard (ME *yerde, yarde,* OE *gerd, gierd* "staff, twig, measuring rod"; Germanic **gazdaz,* IE **ghasto-* "rod, staff") The fundamental unit of length in the Customary System, equal to 3 **feet** or 36 **inches**. Until recently, the yard was defined in England as the distance between two marks on a standard kept at the Tower of London, whereas in the U.S. it was defined as $\frac{3600}{3937}$ **meters**. In 1959, a compromise was arrived at putting the "international yard" at 0.9144 **meters** exactly; however, this yard has been replaced in England by the meter, and has rarely been used in the U.S. for land measure. The corresponding simplified definition of the inch, on the other hand (inch = 2.54 cm) appears to have been adopted generally for very precise calculations involving short lengths. In calculating metric equivalents of land measures, therefore, the original U.S. definition of the yard ($\frac{3600}{3937}$ meters) has been used in this book, whereas in calculating metric equivalents of capacities, the newer definition of the inch (2.54 cm) has been used.

Frequently referenced sources

To save space in footnotes, the following sources are referred to by the name in ALL CAPS. Because of the large number of citations, these sources are not referenced in the Index.

ADAMS, J. Q. *Report upon Weights and Measures, by John Quincy Adams, Secretary of State of the United States.* Washington: Gales & Seaton, 1821.

CLARKE, F. W. *Weights, Measures, and Money, of All Nations.* New York: D. Appleton & Company, 1891.

CONNOR, R. D. *The Weights and Measures of England.* London: Her Majesty's Stationery Office, 1987.

JOHNSON'S *(Revised) Universal Cyclopaedia: A Scientific and Popular Treasury of Useful Knowledge.* S.v. "Weights and Measures." New York: A. J. Johnson Co., 1886.

MARTINI, A. *Manuale di metrologia, ossia misure, pesi e monete in uso attualmente e anticamente presso tutti i popoli.* Torino: Loescher, 1883.

SKINNER, F. G. *Weights and Measures: Their Ancient Origins and Their Development in Great Britain up to AD 1855.* London: Her Majesty's Stationery Office, 1967.

The following abbreviations are also used:

TATE'S refers to Spalding, W. F., ed. *Tate's Modern Cambist,* 27th Edition. London: Effingham Wilson, 1926.

MED refers to the *Middle English Dictionary,* published under various editors in editions from 1952 to 1999 (Ann Arbor: University of Michigan Press).

OED refers to the *Oxford English Dictionary,* 2d Edition (1989), in the electronic version from Oxford University Press.

UN (1955) and UN (1966) refer to the 1955 and 1966 editions of *World Weights and Measures: Handbook for Statisticians,* published by the Statistical Office of the United Nations.

Works cited in the text

The Acts and Resolves, Public and Private, of the Province of the Massachusetts Bay. Boston: Wright & Potter, Printers to the State, 1869.

Acts of Assembly, Passed in the Province of New-York, From 1691, to 1718. London: John Baskett, Printer to the Kings most Excellent Majesty, 1719.

Adams, J. Q. *Report upon Weights and Measures, by John Quincy Adams, Secretary of State of the United States.* Washington: Gales & Seaton, 1821.

Aiken, J. G. *Digest of the Laws of the State of Alabama: Containing all the Statutes of a Public and General Nature, in Force at the Close of the Session of the General Assembly, in January 1833...and Also a Supplement Containing the Public Acts for the Years 1833, 1834, and 1835.* 2d ed. Tuscaloosa (D. Woodruff) and Mobile (Sidney Smith), 1836.

Alexander, J. *A Collection of the British Statutes in Force in Maryland, According to the Report Thereof Made to the General Assembly by the Late Chancellor Kilty.* Baltimore: Cushings & Bailey, 1870.

"An Account of a Comparison lately made by some Gentlemen of the Royal Society, of the Standard of a Yard, and the several Weights lately made for their Use; with the Original Standards of Measures and Weights in the Exchequer, and some other kept for public Use, at Guild-hall, Founders-hall, the Tower, &c." *Philosophical Transactions* 42, no. 470 (April 21 to June 23, 1743): 541–56. New York: Johnson/Kraus Reprint, 1963.

Anderson, R. C., ed. *The Assize of Bread Book, 1477–1517.* Publications of the Southampton Record Society. Southampton: Cox & Sharland, 1923.

The Assise of Bread. Newly corrected & enlarged, from twelue pence the quarter of wheat, vnto three poūd & six pence the quarter according to the rising and falling of the price thereof in the market, by sixe pence altering in euery quarter of wheate, together with sundrie good and needefull ordināces for Bakers, Brewers, Inholders, Victuallers, Vintners, and Butchers: And also other Assises in weightes and measures, which by the lawes of this Realm are commanded to be obserued and kept by all manner of persons, as wel within liberties as without. London: Iohn Winder, 1601.

Bakka, E. "Two Aurar of Gold: Contributions to the Weight History of the Migration Period." *Antiquaries Journal* 58 (1978): 279–98.

Barlow, W. "An Account of the Analogy betwixt English Weights and Measures of Capacity." *Philosophical Transactions* 41, no. 458 (Sept.–Dec. 1740): 457–59.

Bass, G. F. "Cape Gelidonya: A Bronze Age Shipwreck." *Transactions of the American Philosophical Society,* n.s., vol. 57, pt. 8 (1967): 44–142.

Beawes, W. *Lex Mercatoria Rediviva: or, the Merchant's Directory,* 6th ed. Dublin: Printed for James Williams, 1773. Accessed through Google Books.

Berriman, A. E. "The Carvoran 'Modius.'" *Archaeologia Aeliana,* 4th ser., 34 (1956): 130–32.

———. *Historical Metrology: A New Analysis of the Archaeological and the Historical Evidence Relating to Weights and Measures.* London: J. M. Dent & Sons, 1953.

Bettess, F. "The Anglo-Saxon Foot: A Computerized Assessment." *Medieval Archaeology* 35 (1991): 44–50.

Blunt, J. *The Shipmaster's Assistant, and Commercial Digest: Containing Information Necessary for Merchants, Owners, and Masters of Ships,* 11th ed. New York: E. and G. W. Blunt, 1864.

Boyd, J. P., ed. *The Papers of Thomas Jefferson* 16 (30 November 1789 to 4 July 1790). Princeton: Princeton University Press, 1961.

Brent, J. "Account of the Society's Researches in the Anglo-Saxon Cemetery at Sarr." *Archaeologia Cantiana* 6 (1866): 157–63.

Brigham, W., ed. *The Compact with the Charter and*

Works cited in the text

Laws of the Colony of New Plymouth. Boston: Dutton and Wentworth, Printers to the State, 1836.

Brough, B. H. *A Treatise on Mine-Surveying,* 5th ed. London: Charles Griffin & Company, 1896. Accessed through Google Books.

Brown, G.A. and H. Wheeler, eds. *The Compiled Statutes of the State of Nebraska, 1881,* 8th ed. Lincoln: State Journal Company, 1897.

Bruce, J., ed. *Calendar of State Papers, Domestic Series, of the Reign of Charles I: 1631–33.* London: Longman, Green, Longman, and Roberts, 1862. Accessed through Google Books.

Bush, B., ed. *Laws of the Royal Colony of New Jersey, 1703–1745.* New Jersey Archives, 3d ser., Vol. 2. Trenton: New Jersey State Library Archives and History Bureau, 1977.

Canadian Parliament. Reports of experimental work conducted at the Central Experimental Farm, Ottawa, Ontario (appendices to annual reports of the Minister of Agriculture published as Sessional Papers, 1891–1911).

Chase, S. P., ed. *The Statutes of Ohio and of the Northwestern Territory, Adopted or Enacted from 1788 to 1833 Inclusive.* Cincinnati: Corey and Fairbank, 1833.

Chisolm, H.W. "On the Science of Weighing and Measuring, and the Standards of Weight and Measure." *Nature* 8 (1873): 268–70, 307–09, 327–29, 367–70, 386–89, 489–91, 552–55.

———. *On the Science of Weighing and Measuring and Standards of Measure and Weight.* London: MacMillan and Co., 1877.

[———] *Seventh Annual Report of the Warden of the Standards on the Proceedings and Business of the Standard Weights and Measures Department of the Board of Trade.* London: Her Majesty's Stationery Office, 1873.

[———] *Tenth Annual Report of the Warden of the Standards on the Proceedings and Business of the Standard Weights and Measures Department of the Board of Trade.* London: Her Majesty's Stationery Office, 1876.

Chitty, J. *Treatise on the Laws of Commerce and Manufactures, and the Contracts Relating Thereto.* London: A. Strahan, Law-Printer to the King's Most Excellent Majesty, 1824. Accessed through Google Books.

Clarke, F. W. *Weights, Measures, and Money, of All Nations.* New York: D. Appleton & Company, 1891. Accessed through Google Books and supplemented in the case of missing pages with online editions of 1875 and 1877 and a print edition of 1885.

The Code of Virginia. Richmond: James E. Goode, 1887.

"The Connection Between Chinese Music, Weights, and Measures" (no author). *Nature* 30 (Oct. 9, 1884): 565–66.

Connor, R. D. *The Weights and Measures of England.* London: Her Majesty's Stationery Office, 1987.

Conversations-Lexikon Brockhaus, 11th ed. S.v. "Scheffel." Leipzig: F. A. Brockhaus, 1864.

Cooper, T. *Two Tracts: On the Proposed Alteration of the Tariff; and on Weights & Measures. Submitted to the Consideration of the Members from South-Carolina, in the Ensuing Congress of 1823–24.* Charleston: A. E. Miller, 1823.

Cunliffe, B. *Europe between the Oceans, 9000 BC – AD 1000.* New Haven: Yale University Press, 2008.

Curry, P. A. *The Control of Weights and Measures in Egypt.* Ministry of Public Works, Egypt. Physical Department Paper no. 34. Cairo: Government Press, 1939.

Davidson, D. *The Hidden Truth in Myth and Ritual and in the Common Culture Pattern of Ancient Metrology,* 2nd ed. London: Covenant Publishing, November 1940.

Dean, J. E. *Epiphanius' Treatise on Weights and Measures: The Syriac Version.* Oriental Institute of the University of Chicago Studies in Ancient Oriental Civilization, no. 11. Chicago: University of Chicago Press, 1935.

Decennial Edition of the American Digest (1906). St. Paul: West Publishing Co., 1910.

Deetz, P., and C. Fennell, eds. *Records of the Colony of New Plymouth in New England.* Plymouth Colony Archive Project, 1999–2007. Source: "*The Records of the Colony of New Plymouth in New England,* 'Printed by Order of the Legislature of the Commonwealth of Massachusetts,' edited by David Pulsifer, Clerk of the Office of the Secretary of the Commonwealth, and printed by The Press of William White, Printer to the Commonwealth, 1861. These records were reprinted in facsimile form as the *Records of the Colony of New Plymouth in New England*, edited by Nathaniel B. Shurtleff and David Pulsifer, and printed by AMS Press, New York, 1968." http://www.histarch.uiuc.edu/plymouth/laws1.html

Drovetti ("M. Le Chevalier"). *Lettre a M. Abe Remusat sur une Nouvelle Mesure de Coudée, vee a Memphis*. Paris: De Bure Freres, 1827.

The Earliest Printed Laws of North Carolina 1699–1751. Vol. 1: Collection of Public Acts in Force. Wilmington, Delaware: Michael Glazier, Inc., 1977.

Encyclopædia Britannica, 11th ed. S.v. "Pheidon."

Encyclopedia of Industrial Chemical Analysis. New York: Interscience, 1974.

Evans, John. *A Popular History of the Ancient Britons or the Welsh People, from the Earliest Times to the End of the Nineteenth Century.* London: Elliot Stock, 1901. Accessed through Google Books.

Eusebio, L. *Compendio di Metrologia Universale e Vocabolario Metrologico*. Torino, 1899 (reprinted 1967 by Forni Editore Bologna).

Ferguson, R. S. "The Carlisle Bushel." *Archaeological Journal* 42 (1885): 303–11.

First Report of the Commissioners Appointed to Consider the Subject of Weights and Measures, 7 July 1819. Sessional Papers of the House of Commons, 1819, vol. 11.

Gardner, M. *Fads and Fallacies in the Name of Science*. New York: Dover Publications, 1957.

General Laws, and Joint Resolutions, Memorials, and Private Acts, Passed at the Third Session of the Legislative Assembly of the Territory of Colorado. Denver: Byers and Dalley, 1864.

The General Shop Book: or, The Tradesman's Universal Director. London: Printed for C. Hitch, J. and J. Rivington, and R. Dodsley, 1753. Accessed through Google Books.

General Statutes of North Carolina. Chapter 81A: Weights and Measures Act of 1975. Raleigh: State of North Carolina Department of Justice, 1985.

George, S., B. Nead, and T. McCamant, eds. *Charter of William Penn and Laws of the Province of Pennsylvania, Passed between the Years 1682 and 1700.* Harrisburg: Lane S. Hart, State Printer, 1879.

Gillings, R. J. *Mathematics in the Time of the Pharaohs.* New York: Dover Publications, 1972.

Glazebrook, R. "Standards of Measurement: Their History and Development." Supplement to *Nature* 128, no. 3218 (July 4, 1931).

Griffith, F. L. "The Metrology of the Medical Papyrus Ebers." *Proceedings of the Society of Biblical Archaeology* 13 (May 5, 1891): 392–406.

———. "Notes on Egyptian Weights and Measures." *Proceedings of the Society of Biblical Archaeology* 14 (June 14, 1892): 403–50.

Grimke, J. *The Public Laws of the State of South-Carolina, from its First Establishment as a British Province down to the Year 1790, Inclusive.* Philadelphia: Aitken & Son, 1790.

Hall, G. E. *Four Leagues of Pecos: A Legal History of the Pecos Grant, 1800–1933.* University of New Mexico Press, 1984. Accessed through Google Books.

Hall, H., and F. J. Nicholas. *Select Tracts and Table Books Relating to English Weights and Measures (1100–1742).* Camden Miscellany 15. Camden 3rd ser., vol. 41, no. 5. London: Offices of the Society, 1929.

Handbook of Chemistry and Physics, 60th ed. Chemical Rubber Company, 1979.

Harris, M. D. *The Coventry Leet Book: or Mayor's Register, Containing the Records of the City Court Leet or View of Frankpledge, A.D. 1420–1555, with Divers Other Matters.* Early English Text Society, Original Series, 135. London: Kegan Paul, Trench, Trübner & Co., and Henry Frowde, Oxford University Press, 1908.

Haverfield, F. "Modius Claytonensis: The Roman Bronze Measure from Carvoran." *Archaeologia Aeliana,* 3rd ser., 13 (1916): 85–102.

Head, B. V. *Historia Numorum: A Manual of Greek Numismatics.* Oxford: At the Clarendon Press, 1887. Accessed through Google Books.

Hemmy, A. S. "The Weight Standards of Ancient Greece and Persia." *Iraq* 5 (1938): 65–81.

Hening, W., ed. *The Statutes at Large; being a Collection of All the Laws of Virginia, from the First Session of the Legislature, in the Year 1619.* Richmond: Samuel Pleasants, Junior, Printer to the Commonwealth, 1809.

Works cited in the text

———. *The Statutes at Large; being a Collection of All the Laws of Virginia, from the First Session of the Legislature, in the Year 1619.* New-York: Printed for the Editor by R. & W. & G. Bartow, 1823.

Henry, R. *The History of Britain,* 4th ed. London: T. Cadell and W. Davies, 1805.

Hill, L. D. *Grain Grades and Standards: Historical Issues Shaping the Future.* Urbana and Chicago: University of Illinois Press, 1990.

Hinz, W. *Islamische Masse und Gewichte umgerechnet ins Metrische System.* Handbuch der Orientalistik, Erg. 1, Heft 1. Leiden: E. J. Brill, 1955.

Hitzl, K. *Die Gewichte griechischer Zeit aus Olympia.* Deutsches Archäologisches Institut: Olympische Forschungen, Band 25. Walter de Gruyter: Berlin and New York, 1996.

Hoang, F. *Notions techniques sur la propriété en Chine avec un choix d'actes e de documents officiels.* Variétés sinologiques no. 11. Chang-hai: Imprimerie de la Mission Catholique, 1897.

Holinshed, R. *Chronicles of England, Scotland, and Irelande.* Vol. 1, bk. 3. London: Imprinted for George Bishop, 1577.

Huckle, J. "A history of Beer and Pubs in Halesworth." http://www.halesworth.ws/museum/pubs/index.php

Huggins, P. J. "Anglo-Saxon Timber Building Measurements: Recent Results." *Medieval Archaeology* 35 (1991): 6–28.

———. "Excavation of Belgic and Romano-British Farm with Middle Saxon Cemetery and Churches at Nazeingbury, Essex, 1975–6." *Essex Archaeology and History* 10 (1978): 64–65.

———. "Yeavering Measurements: An Alternative View." *Medieval Archaeology* 25 (1981): 150–53.

Humphries, H. N. *The Coin Collector's Manual: Comprising an Historical and Critical Account of the Origin and Progress of Coinage, from the Earliest Period to the Fall of the Roman Empire.* London: George Bell and Sons, 1897. Accessed through Google Books.

Hunt, T. F. *The Cereals in America.* New York: Orange Judd Company, 1912.

Hussey, R. *An Essay on the Ancient Weights and Money, and the Roman and Greek Liquid Measures, with an Appendix on the Roman and Greek Foot.* Oxford (for the University), 1836.

Jensen, N. F. "Limits to Growth in Global Food Production." *Science* 201 (28 July 1978): 317–20.

The Jewish Encyclopedia. New York: Ktav Publishing, 1964.

Johnson's (Revised) Universal Cyclopaedia: A Scientific and Popular Treasury of Useful Knowledge, s.v. "Weights and Measures." New York: A. J. Johnson Co., 1886.

Judson, L. V. *Weights and Measures Standards of the United States: A Brief History.* U.S. Department of Commerce, National Bureau of Standards. NBS Special Publication 447. Washington: U.S. Government Printing Office. Originally published October 1963; updated March 1976.

Kaplan, S. L. *Provisioning Paris: Merchants and Millers in the Grain and Flour Trade During the Eighteenth Century.* Ithaca and London: Cornell University Press, 1984.

Keith, Skene. "Observations on the Final Report of the Commissioners of Weights and Measures." *Edinburgh Philosophical Journal* 6 (1822): 41–47.

Kennelly, A. E. *Vestiges of Pre-metric Weights and Measures Persisting in Metric-system Europe, 1926–1927.* New York: The Macmillan Company, 1928.

Kilty, W. (Chancellor of Maryland). *A Report of All Such English Statutes as Existed at the Time of the First Emigration of the People of Maryland, and which by Experience Have Been Found Applicable to their Local and Other Circumstances; and of Such Others As Have Since Been Made in England or Great-Britain, and Have Been Introduced, Used and Practised, by the Courts of Law or Equity; and Also All Such Parts of the Same As May Be Proper to Be Introduced and Incorporated into the Body of the Statute Law of the State. Published under the Directions of the Governor and Council, Pursuant to a Resolution of the General Assembly.* Annapolis: Jehu Chandler, 1811.

Kisch, Bruno. *Scales and Weights: A Historical Outline.* New Haven: Yale University Press, 1965.

Landels, J. G. *Engineering in the Ancient World.* Berkeley: University of California Press, 1978.

Laws of the State of Maine. Brunswick: J. Griffin, for the State, 1821.

Laws of the State of Vermont; Revised and passed by the Legislature, in the year of our Lord, One Thousand Seven Hundred and Ninety Seven. Rutland: Josiah Fay, 1797.

Leaming, A. and J. Spicer, *The Grants, Concessions, and Original Constitutions of the Province of New Jersey.* Philadelphia: W. Bradford, Printer to the King's Most Excellent Majesty for the Province of New Jersey, n.d. (thought to be 1751 or 1752). Reprinted as the "Second Edition" by Honeyman & Company, Somerville, New Jersey, in 1881.

Lewy, H. "Assyro-Babylonian and Isrealite Measures of Capacity and Rates of Seeding." *Journal of the American Oriental Society* 64 (Apr.–Jun. 1944): 65–73.

Lincoln, C. Z., W. H. Johnson, and A. J. Northrup, eds. *The Colonial Laws of New York from the Year 1664 to the Revolution.* Albany: James B. Lyon, State Printer, 1894. Accessed through Google Books.

Littell, W., ed. *The Statute Law of Kentucky.* Frankfort: William Hunter, 1810.

Linn, J. B. *Charter of William Penn and Laws of the Province of Pennsylvania, Passed between the Years 1682 and 1700, Preceded by Duke of York's Laws in Force from the Year 1676 to the Year 1682.* Compiled and Edited by G. Staughton, B. M. Nead, and T. McCamant. Harrisburg: Lane S. Hart, State Printer, 1879.

Littell, W., and J. Swigert. *A Digest of the Statute Law of Kentucky.* Frankfort: Kendall and Russell, 1822.

Löweneck, M. *Peri Didaxeon, Eine Sammlung von Rezepten in Englischer Sprache aus dem 11./12 Jahrhundert, nach einer Handschrift des Britischen Museums.* Erlangen: Verlag von Fr. Junge, 1896. Accessed through Google Books.

Lubbock, John (Lord Avebury). *Pre-historic Times, as Illustrated by Ancient Remains and the Manners and Customs of Modern Savages.* 6th ed. New York: D. Appleton and Co., 1900. Accessed through Google Books.

Lucas, R. B. "On 'The British Standards.'" *Transactions and Proceedings and Report of the Royal Society of South Australia* 9 (1885–6): 18–38.

Mackay, E. J. H. *Further Excavations at Mohenjo-daro: Being an official account of Archaeological Excavations at Mohenjo-daro carried out by the Government of India between the years 1927 and 1931,* vols. 1 and 2. New Delhi: Government of India Press, 1938.

Markham, C. A., ed. *The Records of the Borough of Northampton.* Published by order of the Corporation of the County Borough of Northamptom, 1898. Accessed through Google Books.

Martin, R. M. *History of the Colonies of the British Empire in the West Indies, South America, North America, Asia, Austral-Asia, Africa, and Europe.* London: Wm. H. Allen and Co. and George Routledge, 1843. Accessed through Google Books.

Martini, A. *Manuale di metrologia, ossia misure, pesi e monete in uso attualmente e anticamente presso tutti i popoli.* Torino: Loescher, 1883.

Memorandum, by the Officer in Charge of the Standards, on the Cubic Contents, Depth, and Diameter of the Standard Bushel Measure, 23 April 1879. House of Commons Sessional Papers 1878–9 vol. 65.

Michailidou, A. *Weight and Value in Pre-coinage Societies: an Introduction.* ΜΕΛΕΤΗΜΑΤΑ 42. Research Centre for Greek and Roman Antiquity, National Hellenic Research Foundation. Athens: Diffusion de Bocard, 2005.

Miller, W. H. "On the Construction of the New Imperial Standard Pound, and its Copies of Platinum; and on the Comparison of the Imperial Standard Pound with the Kilogramme des Archives." *Philosophical Transactions* 146 (1856): 753–62.

Morehead, C. S., and M. Brown. *A Digest of the Statute Laws of Kentucky.* Frankfort: Albert G. Hodges, 1834.

Murray, John (publisher). *Handbook of the Bombay Presidency, with an Account of Bombay City,* 2nd ed. London: John Murray, 1881. Accessed through Google Books.

Myers Grosses Konversations-Lexikon, 6th ed., s.v. "Scheffel." Leipzig and Vienna: Bibliographisches Institut, 1902.

Needham, J. *Science and Civilization in China,* vol. 3. Cambridge: Cambridge University Press, 1959.

Neugebauer, O., and A. Sachs. *Mathematical Cuneiform Texts.* New Haven: American Oriental Society, 1986 (originally published 1945).

Nichols, J. L. *The Business Guide, or Safe Methods of Business.* Naperville: J. L. Nichols & Company, 1911.

Nicholson, E. *Men and Measures: A History of Weights and Measures Ancient and Modern.* London: Smith, Elder & Co., 1912.

O'Reilly, J. P. "On Two Jade-handled Brushes." *Nature* 35 (Feb. 3, 1887): 318–19.

Parise, N. "Mina di Ugarit, mina di Karkemish, mina di Khatti." *Dialoghi di Archeologia,* n.s., no. 3 (1981): 155–60.

———. "Unita ponderali e rapporti di cambio nella Siria del Nord." In *Circulation of Goods in Non-palatial Context in the Ancient Near East: Proceedings of the International Conference organized by the Istituto per gli studi Micenei ed Egeo-Anatolici,* A. Archi, ed., 125–58. Roma: Edizioni dell'ateneo, 1984.

Parker, R. A. *Demotic Mathematical Papyri.* Providence: Brown University Press, 1972.

Pennick, N. *Natural Measure.* Cambridge: Runestaff Publications, 1985.

Peet, T. E. "Mathematics in Ancient Egypt." *Bulletin of the John Rylands Library, Manchester* vol. 15 no. 2 (July 1931): 409–41.

Petrie, W. M. F. *Ancient Weights and Measures: Illustrated by the Egyptian Collection in University College, London.* London: Department of Egyptology, University College, 1926.

———. *Inductive Metrology; or, the Recovery of Ancient Measures from the Monuments.* London: Hargrove Saunders, 1877.

———. *Measures and Weights.* London: Methuen & Co., 1934.

———. "Measures and Weights (Ancient)." *Encyclopædia Britannica,* 14th ed.

———. *Prehistoric Egypt.* Publications of the Egyptian Research Account and British School of Archaeology in Egypt, 31. London: British School of Archaeology in Egypt, 1920.

———. *The Pyramids and Temples of Gizeh.* London: Field & Tuer [et al.], 1883.

———. "Report of Diggings in Silbury Hill, August, 1922." *Wiltshire Archaeological & Natural History Magazine* 42 (Dec. 1922–Dec. 1924) no. 138, 215–18.

———. Review of Quibell (1913) in *Ancient Egypt* (1915), Part I, 37–41.

———. "Weights and Measures (Ancient Historical)." *Encyclopædia Britannica,* 11th ed.

Phillips, U. B. *The Revised Statutes of Louisiana.* New Orleans: John Claiborne, State Printer, 1856.

Pickering, D. *The Statutes at Large.* Cambridge: Joseph Bentham, Printer to the University, 1762–75 (in part).

Postgate, J. N. "An Inscribed Jar from Tell al Rimah." *Iraq* 40, pt. 1 (Spring 1978): 71–75.

Potter, H. *Laws of the State of North-Carolina.* Raleigh: J. Galls, 1821.

Prince, O. *A Digest of the Laws of the State of Georgia, Second Edition.* Athens: by the Author under the authority of the General Assembly, 1837.

Prinsep, J., ed. "On the Standard Weights of England and India" (no author). *Journal of the Asiatic Society of Bengal,* no. 10 (October 1832): 442–49. Calcutta: Baptist Mission Press, 1832. Accessed through Google Books.

Proceedings of the Society of Antiquaries of London, 2nd ser., 16 (1895–97): 174–75.

The Public Statute Laws of the State of Connecticut, bk. 1. Published by authority of the General Assembly. Hartford: Hudson and Goodwin, 1808.

Pulak, C. "The balance weights from the Late Bronze Age shipwreck at Uluburun." In *Metals Make the World Go Round: The Supply and Circulation of Metals in Bronze Age Europe.* Proceedings of a conference held at the University of Birmingham in June 1997. C. F. E. Pare, ed. Oxford: Oxbow Books, 2000.

Quibell, J. E. *Excavations at Saqqara (1911–12): The Tomb of Hesy.* Cairo: Imprimerie de l'Institut Français d'Archéologie Orientale, 1913.

Ray, P., ed. *Tap's Arithmetick, or, the Path-Way to the Knowledge of the Ground of Arts.* 2d ed. London: J. Streater, 1658.

Redington, J., ed. "Volume 73: March 1–April 29, 1701," *Calendar of Treasury Papers, Volume 2: 1697–1702 (1871),* 467–87. http://www.british-history.ac.uk/report.aspx?compid=79548

Report by the Board of Trade on their Proceedings and Business under the Weights and Measures Acts,

Works cited in the text

1909. London: His Majesty's Stationery Office, 1909.

Report by the Board of Trade on their Proceedings and Business under the Weights and Measures Acts, 1910. London: His Majesty's Stationery Office, 1910.

A Report from the Committee, Appointed to Enquire into the Original Standards of Weights and Measures in this Kingdom, and to consider the Laws relating thereto, with the Proceedings of the House thereupon. London: By Order of the House of Commons, 1758. Cited in the text as "Report from the Committee (1758)." Cf. next entry.

A Report from the Committee, Appointed (Upon the 1st Day of December 1758) to Enquire into the Original Standards of Weights and Measures in this Kingdom, and to consider the Laws relating thereto, with the Proceedings of the House thereupon. London: By Order of the House of Commons, 1759. Cited in the text as "Report from the Committee (1759)." Cf. previous entry.

Report from the Select Committee of the House of Lords appointed to consider the Petition of the Directors of the Chamber of Commerce and Manufactures, established by Royal Charter in the City of Glasgow, 1823. Sessional Papers of the House of Commons 1824, vol. 7.

Report from the Select Committee on the Sale of Corn, 25 July 1834. Sessional Papers of the House of Commons 1834, vol. 7.

Report from the Select Committee on Weights and Measures, 28 May 1821. Sessional Papers of the House of Commons 1821, vol. 4.

Report of the Commissioners appointed to consider the Steps to be taken for Restoration of the Standards of Weight and Measure, 1841. Sessional Papers of the House of Commons 1854–55, vol. 15.

Report of the Select Committee Appointed to Enquire into the Original Standards of Weights and Measures in this Kingdom. Sessional Papers of the House of Commons 1813–1814, vol. 3.

The Revised Code of Laws of Illinois. Shawneetown: Alexander F. Grant & Co., 1829.

The Revised Code of the Laws of Virginia: Being a Collection of All Such Acts of the General Assembly, of a Public and Permanent Nature, as are Now in Force. Richmond: Thomas Ritchie, Printer to the Commonwealth, 1819.

The Revised Statutes of the State of Maine, 2nd ed. Hallowell: Glazier, Masters & Smith, 1847.

The Revised Statutes of the State of New-York, As Altered by the Legislature; Including the Statutory Provisions of a General Nature, Passed from 1828 to 1835 Inclusive. Albany: Packard and Van Benthuysen, 1836.

The Revised Statutes of the State of New-York, Passed During the Years One Thousand Eight Hundred and Twenty-Seven, and One Thousand Eight Hundred and Twenty-Eight. Albany: Packard and Van Benthuysen, 1829.

Reynardson, S. "A State of the English Weights and Measures of Capacity, as they appear from the Laws as well ancient and modern." *Philosophical Transactions* 46 no. 491 (Jan.–Mar. 1749): 54–71. New York: Johnson/Kraus Reprint, 1963.

Robertson, A. J. *The Laws of the Kings of England from Edmund to Henry I.* Cambridge: At the University Press, 1925.

Robson, Eleanor. "Mesopotamian Mathematics." In *The Mathematics of Egypt, Mesopotamia, China, India, and Islam: A Sourcebook,* V. J. Katz, ed., 57–185. Princeton: Princeton University Press, 2007. Accessed through Google Books.

"Roman Libra, with Greek inscription, from North Lincolnshire." *Antiquaries Journal* 13 (1933): 57–58.

Rotuli Parliamentorum; ut et petitiones, et placita in Parliamento. Great Britain. Parliament. Vol. 5. [London], [1767–77].

Scheil, V. and L. Legrain. *Textes Élamites-Sémitiques,* 5th ser. Memoires de la Mission Archéologique de Susiane 14. Paris: Ernest Leroux, 1913.

Schilbach, J. "Massbecher aus Olympia." In *XI. Bericht über die Ausgrabungen in Olympia,* A. Mallwitz, ed., 323–56. Deutsches Archäologisches Institut. Berlin: de Gruyter, 1999.

Schmidt, E. F. *Persepolis II.* University of Chicago Oriental Institute Publications 69. Chicago: University of Chicago Press, 1957.

Scott, E. *Laws of the State of Tennessee, Including those of North Carolina Now in Force in this State, from the Year 1715 to the Year 1820, Inclusive.* Knoxville: Heiskell & Brown, 1821.

Works cited in the text

Scott, R. B. Y. "Weights and Measures of the Bible." *Biblical Archaeologist* 22 no. 2 (1959): 22–40.

Second Decennial Edition of the American Digest (1916). St. Paul: West Publishing Co., 1922.

Skinner, F. G. *Weights and Measures: Their Ancient Origins and Their Development in Great Britain up to AD 1855.* London: Her Majesty's Stationery Office, 1967.

Skinner, F. G. and R. L. S. Bruce-Mitford. "A Celtic Balance-beam of the Christian Period." *Antiquaries Journal* 20 (1940): 87–102.

Sloley, R. W. "Ancient Egyptian Mathematics." *Ancient Egypt* (1922) pt. 4, 111–17.

Smith, A. *An Inquiry into the Nature and Causes of the Wealth of Nations.* Edinburgh: Oliphant, Waugh & Innes, 1814. Accessed through Google Books.

Smith, R. A. "Early Anglo-Saxon Weights." *Antiquaries Journal* 3 (1923): 122–29.

Smith, W. *Dictionary of Greek and Roman Antiquities,* 2d ed. London: Walton and Maberly, 1853. Accessed through Google Books.

Spalding, W. F., ed. *Tate's Modern Cambist,* 27th ed. London: Effingham Wilson, 1926.

Stratmann, F. H. *A Middle-English Dictionary: Containing Words Used by English Writers from the Twelfth to the Fifteenth Century.* A new edition, re-arranged, revised, and enlarged by Henry Bradley. Oxford: At the Clarendon Press, 1891.

The Statutes at Large of Pennsylvania from 1682 to 1801. Compiled under the Authority of the Act of May 19 1887 by James T. Mitchell and Henry Flanders, Commissioners.

Statutes of the Mississippi Territory. Digested by the authority of the General Assembly. Natchez: Peter Isler, Printer to the Territory, 1816.

The Statutes of the Realm: Printed by Command of His Majesty King George the Third in Pursuance of an Address of the House of Commons of Great Britain: From Original Records and Authentic Manuscripts. Published in 11 volumes from 1810 to 1828.

Swan, J. R., ed. *Statutes of the State of Ohio, of a General Nature.* Columbus: Samuel Medary, State Printer, 1841.

Third Decennial Edition of the American Digest (1926). St. Paul: West Publishing Co., 1929.

Thom, A. "The Geometry of Cup-and-Ring Marks." *Transactions of the Ancient Monuments Society,* n.s., vol. 16 (1969): 77–87.

———. *Megalithic Sites in Britain.* Oxford: Clarendon Press, 1967.

Thom, A., and A. S. Thom. *Megalithic Remains in Britain and Brittany.* Oxford: Clarendon Press, 1978.

Thureau-Dangin, F. "La Mesure du *qa*." *Revue d'Assyriologie et d'Archéologie Orientale* 9 (1912): 24–25.

———. "L'u, le Qa et la Mine, Leur Mesure et Leur Rapport." *Journal Asiatique,* 10th ser., 13 (1909): 79–111.

———. "Numération et Métrologie Sumériennes." *Revue d'Assyriologie et d'Archeologie Orientale* 18 no. 3 (1921): 123–42.

———. "Sketch of a History of the Sexagesimal System." *Osiris* 7 (1939): 95–141.

Toulmin, H., ed. *Digest of the Laws of the State of Alabama.* Cahawba: Ginn & Curtis, 1823.

[UN] *World Weights and Measures: Handbook for Statisticians.* Series M No. 21 (Provisional Edition). Statistical Office of the United Nations in Collaboration with the Food and Agriculture Organization of the United Nations. New York: United Nations, 1955.

[UN] *World Weights and Measures: Handbook for Statisticians.* Series M No. 21 Rev. 1. Statistical Office of the United Nations. New York: United Nations, 1966.

[UN] *Glossary of Commodity Terms: Including Currencies, Weights and Measures Used in Certain Countries of Asia and the Far East.* E/CN.11/394. United Nations Economic Commission for Asia and the Far East. United Nations: Department of Economic Affairs, November 1954.

U.S. Department of Agriculture (USDA) Agricultural Marketing Service, Grain Division. *Grain Inspection Manual: Covering the Sampling, Inspection, Grading and Certification of Grain under the United States Grain Standards Act as Amended.* Instruction No. 918(GR)–6). Hyattsville, 1977.

U.S. Department of Agriculture (USDA). *Weather, Crops, and Markets,* vol. 1 no. 11 (March 1922) through vol. 8 no. 11 (November 1931). Washington: U.S. Government Printing Office.

U.S. Department of Commerce, National Bureau of Standards. *A Metric America: A Decision Whose Time Has Come.* U.S. National Bureau of Standards Special Publication 345. Washington: U.S. Government Printing Office, July 1971.

U.S. Department of Commerce, National Bureau of Standards. *Units and Systems of Weights and Measures: Their Origin, Development, and Present Status.* Letter Circular LC 1035. Gaithersburg, November 1985.

U.S. Congress. House. Committee on Coinage, Weights, and Measures. *Standard Barrel for Dry Commodities.* 62d Congress, 2d Session, 1912. Committee Report No. 1120 (to accompany H.R. 23113).

[University of Nebraska, Agricultural Experiment Station.] *Results of the International Winter Wheat Performance Nursery.* Research Bulletins nos. 245 (July 1971) through 305 (March 1984) of the Nebraska Agricultural Research Station.

Wright, J. *The English Dialect Dictionary.* New York: G. P. Putnam's Sons, 1904.

Wylie, C. C. "On King Solomon's Molten Sea." *Biblical Archaeologist,* vol. 12 no. 4 (1949): 86–90.

York Memorandum Book, pt. 2 (1388–1493), Lettered $\frac{A}{Y}$ in the Guildhall Muniment Room. Publications of the Surtees Society 125. Durham and London: Published for the Society, 1915.

Zambaur, E. V. S. v. "Dirham." *E. J. Brill's First Encyclopædia of Islam, 1913–1936* 2:978–79. Accessed through Google Books.

Zuidhof, A. "King Solomon's Molten Sea and (π)." *Biblical Archeologist,* Summer 1982, 179–84.

Zupko, R. E. *British Weights & Measures: A History from Antiquity to the Seventeenth Century.* Madison: University of Wisconsin Press, 1977.

Index

A Metric America, 107
Abbassi, *see* Arabic weights
Acre, 4–12, 106, 131, 139
Acre-breadth (chain), 4, 139
Acts and Resolves of the Province of the Massachusetts Bay, 64
Acts of Assembly, Passed in the Province of New-York (1719), 12, 64
Adams, J. Q., *Report to Congress on Weights and Measures*, ix, 1, 18, 74, 77, 96, 97, 101
Aegina, 24
Aeginetan talent, 25, 35, 36, 78
Æthelstan, 84
Afghanistan, 9, 33
Aiken, J. G., *Digest of the Laws of the State of Alabama*, 94
Airy, W., *Minutes of Proc. Inst. of Civil Engineers*, 39
Alabama, 56, 94
Alaska, 64
Ale, 44, 69, 70, 102, 123, 131, 135–137
Ale gallon, *see* Gallon
Aleppo, 32
Alexander, J., *Collection of the British Statutes in Force in Maryland* (1870), 138
Alexandria, 31, 35
Algiers, 31
ΑΛ[Φ]Ι (αλφιτον), 62
Alum, 26
Amphora, 63
Amsterdam, 101
"An Account of a Comparison lately made by some Gentlemen of the Royal Society", 104
Ancient weights
    Aeginetan, 24
    Argive, 23
    Didrachma, 20, 141

Drachma, 19, 20, 24, 30, 74–76, 141
    Aeginetan, 24
    Attic, 74, 75
Egyptian, *see* Egyptian weights
Etruscan, 24
Mina, 20, 144
    Aeginetan, 24
    Attic, 74
    Sumerian, 20
    Syrian, 23, 29, 77
Seed, 20
Shekel, 20, 147
    Attic, 147
    Egyptian, 21–26, 29, 30, 35, 75, 77, 79, 147
    Hittite, 23
    Palestinian/Hebrew, 23, 147
    Sumerian, 20, 85, 147
Stater, 147
    Argive, 23, 24
    Argive coin, 23
    Attic, 74, 75, 77
Talent, 20, 148
    Aeginetan, 24, 25, 35, 36, 78
    Attic, 74
    Syrian, 23, 35, 75, 78
Tetradrachma, 84, 148
Anglo-Egyptian Sudan, 9
Anglo-Israel movement, vii
Anker, 49, 53, 54
Apothecaries' weight
    Customary, 124
        Dram, 19, 29, 141
        Scruple, 19
    Denmark, 25
    Germany, 25
    Nuremburg, 25
    Russia, 25
Arabic troy cognates

Index

19th century, 31–32
20th century, 33–34
Arabic weights
    Abbassi, 33
    Bahar, 35
    Coffala, 31
    Cossera, 76
    Danar, 33
    Dinar, 35
    Dirhem, 30–34, 76
        Gold, 25
        Silver, 84
    Frehsil, 35
    Kantar, 35
    Maqar, 76
    Maund, 31, 35
    Miscal (metical), 30–33
    Oka, 35
    Pinar, 33
    Rotl (ratl, rottolo, etc.), 31, 33, 76
    Rotl-feudi, 31
    Ruba, 35
    Ukiah (okia, vachia, etc.), 31–33
        Gold, 25
    Woket, 33
Archine (Russia), 100
Are, 11, 106, 139
Argive weight system of Pheidon, 23
    Stater, 23, 24
Argos, 23
Arkansas, 64
Aroba (Peru), 59
Arrack, 77
Asian weights
    Catty, 36
        North Vietnam, 37
        Sumatra and Java, 37
    Hiyaka-me (Japan), 38
    Kwan (Japan), 38
    Mace, 36
    Pecul (Siam), 37
    Pong (Laos), 38
    Tael or liang, 36–78
Assisa Panis et Cervisie (Assize of Bread and Ale, 51 Henry 3, 1266), 71, 111, 122–127, 131
Assize of Bread Book, 90
Assize of London for casks, 44, 46, 47, 98

Assize of the tun, 48, 132
Athens, 74
Aune, 8, 9, 141
Aurar (Viking), 76
Australian Wheat Board, 127
Austro-Hungarian Empire, 30, 54, 75, 84
Average bulk weights of wheat and barley, 111–127
Average test weights of U.S. wheat and barley, 112–114
Avoirdupois (meaning), 87, 89–91, 131–133
Avoirdupois weight, 70, 87–93, 135
    Clove, 88, 92, 107, 134, 140, 145
    Dram, 16, 88, 107, 141
    Half hundred, 88
    Hundredweight, 88, 123, 141, 143
    Ounce, 16–18, 70, 88, 100, 107
        Definition, 88, 93
    Pound, 15–18, 26, 29, 41, 87, 101, 107, 124, 134
        Decimal, 93
        Definition, 16, 18, 88, 93
        Exchequer standard of Elizabeth I (bell form), 89
        Exchequer standard of Elizabeth I (disc form), 92
    Quarter, 88, 146
    Sack, 88, 92, 147
    Sarpler, 90, 147
    Stone, 88, 92, 147
    Tod, 92, 148
    Ton, 88, 107, 148

Babylon, 62
Babylonian measures, *see* Mesopotamian measures
Babylonian Talmud, 63
Baden-Württemberg, 13
Bag of grain or salt, 136
Bahar, *see* Arabic weights
Bakka, E., "Two Aurar of Gold", 85
Baltic Sea, 49
Barbrow, Louis E., xi
Barilla company, 120
Barley, 19, 20, 45, 62, 70, 96, 111–127
    Test weight, 111–127
        Cornell, 117–118
        Generic, 112–114

Six-row vs. two-row, 118
  Specific varieties, 117–118
Barley density vs. water density, 118
Barley density vs. wheat density, 77–78, 111, 113, 114, 117
Barleycorn (unit of length), 103, 131
Barlow, W., "An Account of the Analogy betwixt English Weights and Measures of Capacity", 104
Barrel, *see* Cask measures
Barrel for cranberries, 69
Barrel for fruits and vegetables, 69
Barrel of flour, 64
Barrel of petroleum, *see* Tierce
Barrique (France), 49
Bass, G. F., "Cape Gelidonya", 38
Batavia, 37
Bath (Hebrew), 62
Bath, Maine, 104
Beal, K. A., xi, 38
Beawes, W., *Lex Mercatoria Rediviva*, 128
Beef, 28, 44, 45, 56, 69, 90, 135
Beer, 44, 69, 70, 102, 123, 135–137
Beer barrel, 45
Beijing, *see* Peking
Beirut, 32
Belgium, 9
Bengal, 76
Beqa, *see* Egyptian weights
Berlin, 101
Berriman, A. E.
  "The Carvoran 'Modius'", 65
  Historical Metrology, ix, 46
Bet-el-faki, 31
Bettess, F., "The Anglo-Saxon Foot", 13
Beverages, 108
Bible references
  2 Chronicles 2:10, 66
  2 Chronicles 4:5, 63
  2 Chronicles 35:20, 38
  Exodus 16:36, 66
  Exodus 29:40, 63
  Ezekiel 40:5 and 43:13, 63
  Ezekiel 45:11 and 45:14, 66
  Ezekiel 45:11, 45:13, and 46:14, 66
  Genesis 18:6, 66
  Isaiah 10:9, 38
  Jeremiah 46:2, 38

  1 Kings 7:26, 63
  1 Kings 18:32, 66
  2 Kings 7:1 and 7:18, 66
  Numbers 15, 63
  1 Samuel 13:19–22, 38
  1 Samuel 14:14, 12
  1 Samuel 25:18, 66
Binary series
  Divisions of the inch, 106, 107, 109
  Divisions of the yard, 103
  Measures, 41–44, 55–57, 59–62, 79
  Weights, 15, 88
Bombay, 77
Bordeaux, 49
Borneo, 37
Bosak-Schroeder, Clara, xi, 84
Bourgogne, 49
Boyd, J. P., *Papers of Thomas Jefferson*, 103, 109
Braccio, 9
Brandy, 124
Brazil, 59
Bread, 15, 29, 108, 123, 124
Bremen, 28, 78
Brent, J., "Anglo-Saxon Cemetery at Sarr", 39
Brewers, 102
Brigham, W., *Compact and Laws of the Colony of New Plymouth*, 82
Bristol wine measures of Henry VII, 80
Brown and Wheeler, *Compiled Statutes of the State of Nebraska*, 82
bu/ac, 127, 140
Bullets, 19
Bush, B., *Laws of the Royal Colony of New Jersey*, 64
Bushel, *see* Winchester measure
  Connecticut/Sumerian, 58
  Imperial, 51, 119, 127, 140
  Of 8 wine gallons, 45, 111, 122
Bushel for coal, 137
Butt, *see* Cask measures
Butter, 123, 136

Cab (Hebrew), 63
Cairo, 31
Calais, 72
Calcutta, 36, 76
California, 9

Index

Cambodia, 38
Canada, 105
Canna of Rome, 10
Canne of Marseilles, 8, 13
Capacity vs. weight, *see* Connection between volume/capacity and weight
Cape foot, *see* Foot, Rhineland
Cape Gelidonya shipwreck, 21
Capsules, 42
Carat, 19
Carchemish, 29
Carvoran measure, 50–53, 63
Cask measures, 44–45
  Barrel, 44, 45, 47, 48, 55, 56, 98–100, 108, 123, 134–136, 139
  Barrel of 32 gallons, 55–57, 122, 135
  Firkin, 44, 48, 55, 56, 122, 134–136, 142
  Half barrel, 45, 48, 55, 134
  Hogshead, 44–48, 98–100, 123, 124, 134, 135, 143
  Kilderkin, 44, 56, 135, 136, 143
  Pipe, 146
  Pipe or butt, 44, 46–48, 98–100, 123, 124, 132, 134, 135, 140
  Puncheon, *see* Tertian
  Rundlet, 47, 48, 123, 134, 135, 147
  Tertian or puncheon, 44, 45, 47, 48, 53, 98–100, 123, 134, 135, 146, 148
  Tierce (petroleum barrel), 44, 45, 48, 53, 98–100, 123, 124, 135, 148
  Tun, 44, 46–48, 98–100, 123, 124, 132, 134, 135, 149
  U.S. definitions, 44
Castlereagh, Lord, 9
Catty or kati, *see* Asian weights
Celts, 76
Centiliter, 105, 140
Centimeter, 106, 140
Ch'ien, *see* Mace
Chain (Customary), 4–12, 140
Chalder for cinders, 137
Chalder or chaldron for coal, 136, 137, 140
Chaldron, *see* Winchester measure
Champagne, 49
Charlemagne, 30, 84
Charles I, 69
Cheese, 107, 131
Chicago Board of Trade, 119, 125, 127

Chih, 8, 10
China, 37, 105
Chinese measures, 8–109
Chinese metrology, 1
Chinese weights, 36
Chisolm, H. W.
  "On the Science of Weighing and Measuring", 16
  On the Science of Weighing and Measuring, 15, 16, 46, 95
  Seventh Annual Report, 15, 25, 28, 30, 71–73, 91
  Tenth Annual Report, 55
Chitty, J., *Treatise on the Laws of Commerce and Manufactures*, 128
Cider, 44, 124
Cinders, 137
City of Bristol Museum, xi
Cloth, 131–134
Cloth guilds, 102
Coal, 135–137
Code of Virginia (1877), 83
Codfish, 56
Coffala, *see* Arabic weights
Cognac, 49
Coins and coinage, 18–20, 24, 27, 29, 30, 35–39, 65, 72, 74, 76, 84, 133, 141, 145, 148
Cologne, 29–30, 73, 75–77, 85, 91
Colorado, 9
Commissioners of the Excise, 46, 47
Common-law basis of the Customary System, 129–130
Compositio Ulnarum et Perticarum (Composition of Yards and Perches, 33 Edward 1 stat. 6, 1301), 5, 130, 131
Connecticut, 58, 112
"The Connection Between Chinese Music, Weights, and Measures", 40
Connection between volume/capacity and weight, 45–46
  32 cubic feet, 101
  Ale gallon, 87
  Cubic English foot, 97, 100–102
  Cubic Greek foot, 101
  Cubic inch, 96–98
  Cubic Northern foot, 54, 77, 97
  Cubic Roman foot, 97

Guildhall gallon, 88
Liter, 93
Roman trade hemina, 52
Six-packs, 93
Syrian/Phoenecian kotyle, 78
U.S. fluid ounce, 78
Winchester bushel, 77, 97, 100
Winchester pint, 77
Wine gallon, 93
Wine pint, 46, 77
Connor, R. D., *The Weights and Measures of England*, ix
Constantinople, 30, 32
Conversations-Lexikon Brockhaus, 85
Cooper, T., *Outlines of a Proposed Act of Congress*, 109
Coopers, 45, 60, 100, 102, 135
Coopers, 80
Coopers guild, 102
Copper, 26
Cornell grain weight data, 116–118
Cornell University, xi
Cosmetics, 43
Courts-leet, 28
Covado, 9
Cranberry barrel, 69
Cubic cubit
    Egyptian, 54, 62
    Greek, 79
    Hebrew, 63
    Sumerian, 56–59, 62
Cubic digit
    Roman, 85
Cubic foot
    And the number 80, 96–98
    English, 97–98, 100–102
    Greek, 54, 101
    Northern, 54, 55, 77, 78
    Roman, 51, 77
    Sumerian, 54, 58–59
Cubic t'sun, 78
Cubit, 95, 140
    Egyptian, 11, 54, 62, 63, 101
    Hebrew, 63
    Northern, 9
    Sumerian, 54
"Cubit and a handsbreadth" (Ezekiel 40:5 and 43:13), 63

Cuddy (Arabia), 59, 63
Cunliffe, B., *Europe between the Oceans*, 65
Cup (Customary), *see* Half pint
Cup weights, 15
Customary measures
    Areas, 4–12
    Dry capacities, 67–82
    Lengths, 3–12
    Liquid capacities, 41–63
    Weights, 15–38, 87–93
Customary vs. metric
    Beer bottle, 106
    Building materials, 107
    Capacity measures for cooking, 105
    Containers for wine and spirits, 105
    Dimensions of manufactured goods, 107
    International trade, 107
    Land measures, 106
    Lengths, 105
    Packaging, 108
    Recommendations, 109
    Shorter lengths, 106
    Weights, 107
    Wine bottle, 106
Cwt, 141
Cylinders, 46, 71, 98–100
Cyprus, 22, 33

Damascus, 30, 32
Danar, *see* Arabic weights
Davidson, D., *The Hidden Truth in Myth and Ritual*, ix
Deben, *see* Egyptian weights
Decennial Edition of the American Digest, 138
Decimal monetary system, 107
Decimal number system, 1
Decimal series, 1
    China, 1, 109
    Divisions of the inch, 107
    Divisions of the pound, 107
    U.S. weights and measures, 107
Decimeter, 12, 141
Decree of Emperor Ferdinand I (1 August 1560), 75
Deetz and Fennel, *Records of the Colony of New Plymouth in New England*, 83
Delaware, 64
Demibushel, *see* Winchester measure

Index

Denarius (Roman coin), 19, 141
Denmark, 25, 28, 49
Densities
    Barley, 77, 96, 111–127
    Water, 51, 100–101
    Wheat, 77–78, 96, 100, 111–127
    Wine, 77–78, 96, 101, 111–127
Diamonds, 76
Didrachma, 141
Digit, 141, 142
    Customary, 65, 95
    Greek, 65
    Roman, 52, 65, 85
Dimensions of manufactured goods, 107
Dinar, *see* Arabic weights
Dirhem, *see* Arabic weights
Domitian, 50
Drachma, *see* Ancient weights
Dram
    Apothecaries', *see* Apothecaries' weight
    Arabic, *see* Dirhem
    Avoirdupois, *see* Avoirdupois weight
Dram equivalent, 88
Dresden, 75, 79
Drovetti, *Lettre a M. Abe Remusat*, 65
Drugs, 15, 107
Drusian foot, 52
Drusus, 52
Dry measures (Customary), *see* Winchester measure
Dry substances, 67
Duke of York, 68
Duodecimal (base-12) system of notation, 106
Dwight and Lloyd Sintering Co. v. American Ore Reclamation Co., 263 F. 315 (1920), 129
Dwt, *see* Pennyweight

Earliest Printed Laws of North Carolina, 64
Edgar the Peacable, 67
Edward I, 5, 8, 74, 89, 101
Edward II, 28
Edward III, 85, 90, 92
Eels, 47, 48, 123, 134, 135
Eggs, volume of, 63
Egypt, 33, 62, 76
Egyptian measures
    Cubic cubit, 54, 62
    Cubit, 11, 54, 63
    Hekat, 62
    Hon, 62
    Tomb of Hesy, 59–62
Egyptian metrology, 1
Egyptian weight system, 21
Egyptian weights
    Beqa, 21, 35, 40, 139
    Deben, 21, 23, 141
    Half sep, 21, 29, 77
    Qedet, 21–25, 29, 35, 75, 146
        In England, 26
    Sep, 21, 35, 147
Electuaries, 29, 73, 74, 91
Elizabeth I, 15
    Jury of 1574, 91
    Proclamation of June 1583, 29
Ell, 9, 68, 95, 141
Empresa Pública de Abastecimento de Cereals (EPAC), 128
Encyclopedia of Industrial Chemical Analysis, 85
Entemena of Lagash, 57, 62
Ephah (Hebrew), 62
Eritrea, 33
Estonia, 49
Ethiopia, 33
Etruscan weights, 24
Eveleigh, D. J., xi, 80
Exchequer verdict of 30 October 1527 (18 Henry 8), 72
Excise to Treasury Correspondence, 64
Experimental Farm at Ottawa, 113–114
Export-quality wheat, 119
Eyrir (Viking), 76

Faat, *see* Winchester measure
Fasteners, 106
Fathom, 95, 142
Ferdelh, *see* Ferlingate
Ferdinand I, 75
Ferlingate, 5, 142
Fermented liquors, 45
Ferruckabad, 76
Fifth (wine measure), 108, 142
    In England, 106
    In U.S., 105

Finger, *see* Digit, *see* Digit
Firewood, 135, 137
Firkin, *see* Cask measures
First Report of the Commissioners Appointed to Consider the Subject of Weights and Measures, 103
Fish, 44, 48, 56
Fjell, K. M., 128
Flax, 26, 123, 137
Florence, 87
Flour, 64
Fluid dram, 42
Fluid ounce, 42, 59
Food and Agriculture Organization of the United Nations, 119
Food packaging, 108
Foot, 142
    Austro-Hungarian, 54
    Chinese (Peking and Shanghai) for land measure, 8, 10
    Customary, 5–12, 65, 95–103, 106, 131
    Drusian, 52
    Greek, 54, 74, 95, 101
    International, 11
    Moscow, 8
    Natural, 3
    Northern, 6–12, 54, 65, 78, 95
    Rhineland, 10, 101
    Roman, 4, 51, 52, 95
    Shorter Saxon, 9
    Sumerian, 54, 58, 59
    Survey, 12
    Syrian, 6
Founders Company, 103
Fractions, calculating with, 53
France, 9, 30, 37, 48, 52, 88, 107
Frehsil, *see* Arabic weights
Freije, P. C., xi
French colonial measures, 69
French weights and measures, 30
    Canne of Marseilles, 8
    Livre (pound), 127
    Setier of Paris, 127
Frontignac, 49
Fruits, 67, 83
Furlong, 4–12, 142
Fuss, *see* Foot

Gallatin, A., 17
Gallon, 142
    Ale, 41, 46, 87, 136
        Exchequer standard of Elizabeth I, 83
    Bristol wine measures, 80
    European cognates, 48–49
    Guildhall, 41, 46, 88
    Imperial, 41, 51, 100, 101
    Irish wine, 74
    Scots, 137
    Winchester, 41, 44, 55, 64
    Wine, 41–63, 67, 68, 96, 101, 123, 124, 136
        Definition, 98, 111, 122–127, 135, 137
        Definition (1496), 45–46
        Definition (1706), 41, 46
        Exchequer standard of Queen Anne, 41, 46
        History in England, 46–48
Gardner, M., *Fads and Fallacies in the Name of Science*, vii
Gaz memar of Afghanistan, 9
Gear and Gear, *An Ancient Bird-Shaped Weight System from Lan Na and Burma*, 40
Gedda, 76
General Laws of the Territory of Colorado (1864), 9
The General Shop Book, 128
Generic wheat and barley, 112–114
Genoa, 59
George, Nead, and McCamant, *Charter to William Penn and Laws of the Province of Pennsylvania*, 64
Georgia, 56
German measures, 10, 19
Germany, 9, 25, 36, 49, 78, 101, 128
Gia (Indochina), 54
Gill, 42, 107, 142
Gillings, R. J., *Mathematics in the Time of the Pharaohs*, 66
Gilton, Kent, 26, 75
Glasgow Chamber of Commerce and Manufactures, 100
Glass weights, 30
Glazebrook, R., "Standards of Measurement", 65
Gold, 19, 26, 29, 76, 107, 123, 124, 133

Index

Goldsmiths, 132
Goldsmiths' Company, 102
Gombetta (Genoa), 59
Grading of grain, 111–127
Gradus (Roman), 3
Grain, 131, 136
Grain (troy unit), *see* Troy weight
Grain weight as a natural standard, 111–127
Gram, 143
Great Law (Pennsylvania), 70
Great Pyramid of Egypt, vii, 54
Greco-Roman metrology, 2, 11
Greece, 62, 74
Greek measures, 8
    Cubic cubit, 79
    Cubic foot, 54, 101
    Digit, 65
    Foot, 54, 74
    M6, 62
Griffith, F. L.
    "Notes on Egyptian Weights and Measures", 66
    "The Metrology of the Medical Papyrus Ebers", 65
Grimke, J., *Public Laws of the State of South-Carolina*, 65
Grove Ferry weight, 36, 78
Grove Ferry, Kent, 26
Gudda, *see* Cuddy
Guildhall, 88
Guildhall gallon, *see* Gallon
Guilds, 95, 102–103
Gunpowder, 19
Gunter, E., 12

H:l-Gumælius, T., 114, 128
Hadrian's Wall, 50
Hake, 56
Halesworth, 69
Half kilo, 107
Half pace
    Customary (military pace), 3
    Sumerian, 58
Half pint (cup), 42, 68
Half stone (avoirdupois), *see* Avoirdupois weight, Clove
Hall and Nicholas, *Tracts and Table Books*, 12, 39, 84, 94, 128

Hall, G. E., *Four Leagues of Pecos*, 13
Hand, 95, 143
Handbook of Chemistry and Physics, 85
Hanseatic League, 29, 75, 84
Harris, M. D., *Coventry Leet Book*, 39
Harun-al-Rashid, 30, 84
Hasa, K., 127
Haverdepois ounce, 28
Haverdepois pound, 26–29
Haverfield, F., "Modius Claytonensis: The Roman Bronze Measure from Carvoran", 65
Heaped measures, 67, 83, 129, 132, 133, 136, 141
Hebrew measures, 62–63
Hectare, 11, 106, 143
Hectoliter, 143
Heidelberg, 9
Heinrich, D., 127
Hekat (Egyptian weight), 62
Helm, R. B., 127
Hemina (Roman measure), 51
Hemmy, A. S., "Weight Standards of Ancient Greece and Persia", 84
Hemp, 26, 90, 135, 137
Hening, W.
    Statutes at Large... of the State of Virginia (1809), 82
    Statutes at Large... of the State of Virginia (1823), 64, 83
Henry I, 11, 95
Henry VII, 92
Henry, R., *History of Britain*, 72, 84, 85, 104
Heritage grain, 113
Herring barrel, *see* Cask measures, Barrel of 32 gallons
Herrings, 47, 55, 123, 134, 135
Hesy, 59–62, 78, 79
Hide (area), 5, 143
Hides, 73
Highways, 4
Hill, L. D., *Grain Grades and Standards*, 127
Hin (Hebrew), 62
Hinz, W., *Islamische Masse und Gewichte*, 39, 84
Historical metrology, vii
Hittite shekel, 23

Hitzl, K., *Gewichte griechischer Zeit aus Olympia*, 39, 84
Hiyaka-me, *see* Asian weights
Hoang, F., *Notions Techniques*, 13, 40
Hodeida, 35
Hoffman, Dr., 128
Hogshead, *see* Cask measures
Holinshed's *Chronicles* (1577), 4, 56, 91
Holinshed, R., 56
Holland, 9
Homer, 24
Hon (Egyptian measure), 62
Honey, 47, 48, 56, 123, 133, 134
Hong Kong, 37
Houses of Parliament, destruction by fire, 16
Huckle, J., "A history of Beer and Pubs in Halesworth", 83
Huggins, P. J.
    "Anglo-Saxon Timber Building Measurements", 103
    "Excavation of Belgic and Romano-British Farm", 12
    "Yeavering Measurements: An Alternative View", 12
Humphries, H. N., *Coin Collector's Manual*, 65
Hungarian measures, 10
Hunt, T. F., *The Cereals in America*, 127
Hussey, R., *Essay on the Ancient Weights and Money*, 39, 65, 66

Idaho, 64
Ilium, 24
Illinois, 64
Imperial bushel, *see* Bushel
Imperial fluid ounce, *see* Fluid ounce
Imperial gallon, *see* Gallon
Imperial pint, *see* Pint
Imperial system, 16
Inch, 143
    Customary, 95–103, 106, 131
    Roman (uncia), 65
India, 59, 76
Indian weights
    Chittack, 37
    Maund, 35–37, 77
    Pao, 35, 37
    Pussaree, 37
    Rupee, 36, 77
    Seer, 35, 37
    Sicca, 37, 76
    Sicca weight, 36
    Surat khandi, 36
    Tola, 35, 76
Indochina, 54
Indonesia, 37
Indus valley civilization, *see* Mohenjodaro
International foot, 11
International Maize and Wheat Improvement Center, 119
International Winter Wheat Performance Nursery, 119–122
Iran, 30, 33
Ireland, 127
Iron, 26
Isle of Man, 75
Italian measures, 10
Italian National Institute of Nutrition, 120
Italy, 9, 30, 37, 87, 120
IWWPN, *see* International Winter Wheat Performance Nursery

James I, 68, 102
James, Duke of York, 122
Japan, 37, 38
Jefferson, T., 100, 103, 107
Jerusalem Talmud, 63
Jewels, 123
Jordan, 34, 127
Judson, L. V., *Weights and Measures Standards of the United States*, 38, 82
Juice bottles, 108

Kanne (Saxony), 49
Kantar, *see* Arabic weights
Kaplan, S. L., *Provisioning Paris*, 127
Kater, H., 18
Keith, S., *Observations on the Final Report of the Commisioners of Weights and Measures*, 103
Kennelly, A. E., *Vestiges of Pre-metric Weights and Measures*, 13
Kentucky, 64, 82, 129
kg/ha, 143
Kilderkin, *see* Cask measures
Kilogram, 41, 45, 93, 107, 144
Kilometer, 11, 144

Index

Kilty, W., *Report of All Such English Statutes as Existed at the Time of the First Emigration* (1811), 138
Kin or chang, *see* Catty
Kisch, B., *Scales and Weights*, 39
Klafter
    Austro-Hungarian Empire, 65
    Heidelberg, 9
Korea, 37, 38
Kotyle, 62, 78
Kung or kong (China), 13

Laces, 26
Lachter of Württemberg, 8
Landels, *Engineering in the Ancient World*, 38
Laos, 38
Laws, *see* Statutes
Laws of the State of Maine (1821), 38
Laws of the State of Vermont, 94
lb/bu (defined), 113, 117, 144
Lead, 26, 103, 123, 131
Leaming and Spicer, *The Grants, Concessions, and Original Constitutions of the Province of New Jersey*, 64
Lebanon, 34
Legalization of metric units, 129
Lewy, H., "Assyro-Babylonian and Isrealite Measures of Capacity", 66
Liang, *see* Tael
Libra, 51
Libra mercatoria, *see* Tower weight, Merchant's pound
Libya, 33, 34
Lincoln cent, 19
Lincoln, Johnson, and Northrup, *Colonial Laws of New York*, 64, 82, 83
Ling, 56
Link, 12, 144
Linn, J. B., *Charter of William Penn and Laws of the Province of Pennsylvania*, 83
Liquor, 43, 44
Liter, 41, 45, 93, 105, 144
Lithuanian measures, 10
Littell and Swigert, *Digest of the Statute Law of Kentucky*, 138
Littell, W., *Statute Law of Kentucky*, 82, 138
Livonia, 49
Livre, *see* Pound, French

London faat, *see* Winchester measure
Loth (Vienna), 84
Louisiana, 69, 85
Lucas, R. B., "On 'The British Standards'", 39
Lyons, 52

Mässchen (Saxony), 79
Macau, 37, 59
Mace, *see* Asian weights
Mackay, E. G. H., *Further Excavations at Mohenjo-daro*, 13
Macon, 49
Madder, 26
Madrid, 9
Magna Carta (9 Henry 3, 1225), 80, 130, 131
Maine, 38
Malaya, 37
Malmsey, 47, 134, 135
Malt, 70, 137
Malt beverages, 45
Manor of Halesworth Minute Book 1698, 83
Marc de la Rochelle, 85
Marc of Troyes, 85
Mark (marc), 29, 144
Mark of Cologne, 29
Mark of Vienna, 75
Markham, C. A., *Records of the Borough of Northampton*, 84
Marseilles, 8
Marshall, J. W., 128
Martin, R. M., *History of the Colonies of the British Empire*, 40
Maryland, 129
Massachusetts, 44, 64, 83
Maund, *see* Arabic weights; Indian weights
McKelvey, E., 38
Meal, 83
Meat, 28, 107
Mecca, 31
Medicine cups, 42
Medicines, 124
Medieval Sourcebook, 84
Megalith yard, 58
Meile, 10
Melandra, Derbyshire, 26, 30
Memorandum on the Standard Bushel Measure, 103
Memphis, 65

Merchant Taylors' Company, 95, 102
Mercia, 35, 84
Merföld of Hungary, 10
Mesopotamia, 57, 62
Mesopotamian measures, 57–62, 118, 127
Meter, 11, 106, 144
Metric vs. customary, *see* Customary vs. metric
Metze (Saxony), 79
Mexican land grants, 9
Michailidou, A., *Weight and Value in Pre-coinage Societies*, 38
Michigan, 64, 69
Mile, 144
    Customary, 4–12
Mile (Customary)
    Definition, 135
    Local variations, 137
Milk, 107, 108
Miller, W. H., "On the Construction of the New Imperial Standard Pound", 39, 84, 85
Millet (thousandth of a pound), 93
Milligram, 144
Milliliter, 64, 144
Millimeter, 106, 144
Mina, *see* Ancient weights
Mina (Genoa), 59
Minim, 42, 145
Minnesota, 64
Minutes of Evidence (1862), 103, 127
Miscal (metical), *see* Arabic weights
Mississippi, 64, 94
Mississippi Territory, 94
Missouri, 64
Mohenjodaro, 78
Moldavia, 10
Molten sea of Solomon, 63
Montana, 64
Morehead and Brown, *Digest of the Statute Laws of Kentucky*, 138
Moscow, 8
Mow (Peking and Shanghai), 8
Murray, J., *Handbook of the Bombay Presidency*, 40
Mutton, 28, 90, 135
Mycenaean Greece, 57
Myers Grosses Konversations-Lexikon, 85

Nail (avoirdupois), *see* Avoirdupois weight, Clove
Nail (length), 95, 103, 145
National Institute of Standards and Technology (NIST), xi, 38
Nebraska, 82
Nebraska Agricultural Research Station, 119–122
Needham, J., *Science and Civilization in China*, 2
Neolithic stone circles, 58
Neterket, 59
Neugebauer and Sachs, *Mathematical Cuneiform Texts*, 2
New Jersey, 45, 68
New Mexico, 9
New Plymouth, 70
New York, 4, 44, 45, 68, 69, 100, 125
    New York City, 112
New York State Agricultural Experiment Station at Cornell University, 116–118
Newton, Isaac, 65
Nichols, J. L., *Business Guide*, 103
Nicholson, E., *Men and Measures*, vii, 8, 73, 101
Nin (Siam), 55
Noosfia (Arabia), 59
North Africa, 9
North Carolina, 45, 56, 64, 69
North Linconshire, 102
North Sea, 49
North Vietnam, 37
Northampton, 74
Northern pace, *see* Pace, Northern
Norway, 28, 49, 76, 128
Norwegian Grain Corporation, 128
Noumbre of Weyghtes, 26

O'Reilly, J. P., "On Two Jade-handled Brushes", 40
Obol (Attic), 74
Offa, 35, 84
Ohio, 82, 101, 125
Oil, 44, 45, 47, 48, 100, 123, 133, 134
Oka, *see* Arabic weights
Oksaam (Livonia), 49
Old Minster, 8
Olive oil, 66

Index

Olympia, 62
"One weight and one measure", 80, 131, 132, 135, 136
Oregon, 64
Ort (Germany and Scandinavia), 29
Ottawa Experimental Farm, 113–114
Ounce, 145
    Arabic, *see* Ukiah
    Avoirdupois, *see* Avoirdupois weight
    Cologne, 76, 85
    Dresden, 75
    Haverdepois, 28
    Imperial, 100
    Paris, 76
    Roman, 76
    Roman (uncia), 52
    Strasburgh, 85
    Tower, *see* Tower weight
    Troy, *see* Troy weight
Ounce for liquids, *see* Fluid ounce
Ozingell, Kent, 26

Pace, 145
    Chinese, 8
    Geometric, 3, 95
    Military, 3, 95
    Northern, 8–10
    Sumerian, 58
Packaging, 108
Pagan use of the Celtic/Saxon tradition, ix
Palm, 95, 145
Palma
    Moldavia, 10
    Rome, 10
Paris, 76, 127
Parise, N.
    "Mina di Ugarit, mina di Karkemish, mina di Khatti", 38
    "Unita ponderali e rapporti di cambio nella Siria del Nord", 38
Parker, R. A., *Demotic Mathematical Papyri*, 65
Pascuela, R. N., 128
Passus (Roman), 3
Patterns
    1000 ounces ($62\frac{1}{2}$ pounds), 100–102
    80 pounds, 77–78, 97
    Ratio of $\frac{4}{5}$ ($\frac{12}{15}$, $\frac{40}{50}$, etc.), 29, 73, 77–78, 96–98, 101, 111–127

Ratio of barley density to wheat density, 77–78, 111, 113, 114, 117
Ratio of wheat density to wine density, 77–78
Pearls, 26, 123
Peck, *see* Winchester measure
Peet, T. E., "Mathematics in Ancient Egypt", 65
Peking, 8
Pennick, N., *Natural Measure*, ix
Pennsylvania, 44, 45, 68, 70
Penny (sterling), *see* Tower weight
Pennyweight (troy), *see* Troy weight
Perch (area), 4–12, 146
Perch (length), *see* Rod, Customary
Persepolis, 57
Persia, 30, 59
Peru, 59
Peter the Great, 100
Petrie, W. M. F., 6, 10, 21, 22, 38, 54, 59, 60, 65
    Ancient Egypt, 66
    Ancient Weights and Measures, 12, 38, 39, 65, 66
    Inductive Metrology, xi, 8, 58, 95
    Measures and Weights, 66
    "Measures and Weights (Ancient)", 66, 85
    Prehistoric Egypt, 38
    Pyramids and Temples of Gizeh, 65
    "Report of Diggings in Silbury Hill", 65
    "Weights and Measures (Ancient Historical)", x, 66
Petroleum barrel, *see* Tierce
Peyem, *see* Palestinian/Hebrew shekel
Pheidon, King of Argos, 23
Philippine Islands, 37
Phillips, U. B., *Revised Statutes of Louisiana*, 83
Phoenecian/Syrian kotyle, 62
Pic, 9
Pickering, D., *Statutes at Large*, 130
Pie (pied, piede), *see* Foot
Pilchards, 136
Pillory, 69
Pinar, *see* Arabic weights
Pint, 146
    Wine, 42, 57, 62, 68, 108, 123, 124

Pinte (French), 49
Pipe, *see* Cask measures
Pitch, 45, 56, 64, 69
Platinum, 19
Plumber's Company, 103
Plymouth Colony, 68
Pole (length), *see* Rod, Customary
Pondus, 20
Pork, 28, 44, 45, 56, 69, 90, 135
Portugal, 9, 59, 128
Postgate, J. N., "An Inscribed Jar from Tell al Rimah", 66
Pot of butter, 136
Potter, H., *Laws of the State of North-Carolina*, 65
Pottle, 42, 146
Pound, 146
    Avoirdupois, *see* Avoirdupois weight
    Cologne, 29–30, 75, 91
    French, 127
    German standards, 75
    Haverdepois, 26–29, 85
    Prussia, 29
    Roman, 51, 102
    Tower, *see* Tower weight
    Troy, *see* Troy weight
    Vienna, 84
Pound sterling, 72
Precious metals, 15, 19, 28, 107
Precious stones, 26, 28, 76, 123
Prince, O., *Digest of the Laws of the State of Georgia*, 65
Prinsep, J., "On the Standard Weights of England and India", 40
Proceedings of the Society of Antiquaries of London, 39
Prussia, 29, 78, 101
Public Statute Laws of the State of Connecticut, 127
Pulak, C., "The balance weights from the Late Bronze Age shipwreck at Uluburun", 38
Puncheon, *see* Cask measures
Purcell-O'Byrne, N., 127
Pyramid inch, vii
Pyramidology, vii
Pyx Chamber (Westminster Abbey), 73

Qa (Sumerian), 57
Qedet, *see* Egyptian weights
Quart, 146
    Ale, 68
        Exchequer standard of Elizabeth I, 94
    Wine, 42, 57, 59, 68, 123, 124
Quarter, *see* Winchester measure; Avoirdupois weight
Quarto (Genoa), 59
Quentchen (quintin), 75
    Germany and Scandinavia, 29
    Vienna, 84
Quibell, J. E., *Tomb of Hesy*, 66

Ragil (Sudan), 9
Ratio between bulk weights of wheat and barley, 77–78, 111, 113, 114, 117
Ratio between weights of wheat and wine, 78, 87
Ratio of $\frac{4}{5}$ ($\frac{12}{15}$, $\frac{40}{50}$, etc.), *see* Patterns
Ray, P., *Tap's Arithmetick*, 128
Recommendations, 109
Recycling carts, 56
Relative densities of barley and wheat, 77–78, 111, 113, 114, 117
Relative densities of wheat and wine, 77–78
Report by the Board of Trade (1909), 85
Report by the Board of Trade (1910), 40
Report from the Committee (1758), 39, 72, 83, 94, 103, 104, 109
Report from the Select Committee (1821), 38
Report from the Select Committee of the House of Lords (1823), 39
Report from the Select Committee on the Sale of Corn, 127
Report of the Commissioners, 1841, 12, 38, 94, 103, 109
Report of the Select Committee Appointed to Enquire into the Original Standards, 103
Reputed quart, *see* Fifth (wine measure)
Revised Statutes of the State of Maine (1847), 38
Revised Statutes of the State of New-York (1829), 103
Revised Statutes of the State of New-York (1836), 103
Reynardson, S., "A State of the English

Index

Weights and Measures of Capacity", 103, 104
Ribbons, 26
Robertson, A. J., *Laws of the Kings of England from Edmund to Henry I*, 130
Rod
    Customary, 3–12, 95, 131, 146
    Northern, 9, 96
    Saxon, 96
Rogaland, 76
Roman Gaul, 52
Roman measures, 19, 50–53
Romanian measures, 10
Ronalsdway, 75
Rood, 4–12, 147
Ropes, 26
Rotl (ratl, rottolo, etc.), *see* Arabic weights
Rotuli Parliamentorum, 84
Roub-kaddah (Egypt), 66
Ruba, *see* Arabic weights
Rundlet, *see* Cask measures
Russia, 25, 36, 100

Sa-nekht, 59
Sack, *see* Avoirdupois weight and Wool weights
SAE sizes, 106
Sagene, 100
Sale of Gas Act (1859), 103
Salmon, 47, 48, 134, 135
Salt, 56, 136, 137
Salt beef, 56
Sarawak, 37
Sarpler, *see* Avoirdupois weight and Wool weights
Sarre, Kent, 75
Saudi Arabia, 35
Sauterne, 49
Saxon structures, 96
Saxon weights, 25–26
Saxons, 8, 9, 30, 36, 74, 75, 78, 93, 95, 101
Saxony (kingdom), 49, 75, 79
Scads, 136
Scheffel (Dresden and Württemburg), 78
Scheil and Legrain, *Textes Élamites-Sémitiques*, 66
Schilbach, J., "Massbecher aus Olympia", 66
Schliemann, H., 24

Schmidt, E. F., *Persepolis II*, 65
Schock, schocken, 1
Schroeder, Bethany, xi
Scotch foot, *see* Foot, Rhineland
Scotch troy pound, 26
Scotland, 101
Scott, E., *Laws of the State of Tennessee*, 64, 94
Scott, R. B. Y., "Weights and Measures of the Bible", 66
Scripulum (Roman weight), 19, 20, 52
Scruple
    Apothecaries', 19, 147
    Roman (scripulum), 52, 147
Scrupulum (Roman coin), 65
Se'ah (Hebrew), 62
Sep, *see* Egyptian weights
Setier of Paris, 127
Sexagesimal (base-60) number system, 1–2, 8
Sextarius (Roman measure), 50
Shaftment, 95, 147
Shanghai, 13
Shekel, *see* Ancient weights
Shilling, 19, 20, 26, 147
Shipmaster's Assistant, 125
Shipwrecks, 29
    Cape Gelidonya, 21
    Uluburun, 22
Shiraz, 31
Shotgun loads, 88
SI, 147
SI units, 11
Siam (Thailand), 37, 55
Silbury burial mound, 58
Silchester, 26
Silks, 26
Silver, 19, 26, 29, 36, 39, 76, 107, 123, 124, 132, 134
Singapore, 37
Six-packs, 93, 109
Skinner and Bruce-Mitford, "A Celtic Balance-beam of the Christian Period", 85
Skinner, F. G.
    Weights and Measures, 13
    Weights and Measures, ix, 20
Sloley, R. W., "Ancient Egyptian Mathematics", 65
Smith, A., *Wealth of Nations*, 84

Smith, R. A., "Early Anglo-Saxon Weights", 39, 85
Smith, W., *Dictionary of Greek and Roman Antiquities*, 65
Smyrna, 32
Smyth, C. P., *Our Inheritance in the Great Pyramid*, vii
Soap, 56, 123, 135
Soda, 107
Somalia, 33
Sorrells, M. E., xi, 116
South Africa, 101
South Carolina, 56
Southampton, 90
Spain, 30
Span, 95, 103, 147
Spanish Armada, 15
Spanish measures, 9
Spirits, 59, 77, 105, 106
Spycery (spice, Speise), 28–29
Square chain, 4
Square furlong, 4
Square link, 12
Square pole, *see* Perch (area)
Square rod, *see* Perch (area)
Stadion, 8
Stanjen of Moldavia, 10
Star Chambre at Winchester, 28
Statens Kornforretning, 128
Stater, *see* Ancient weights
Statutes
    American
        Alabama, 56
        Colorado, 9
        Connecticut, 58, 112
        Georgia, 56
        Kentucky, 82, 129
        Louisiana, 69, 85
        Maine, 38
        Maryland, 129
        Massachusetts, 44, 83
        Michigan, 69
        Mississippi Territory, 94
        Nebraska, 82
        New Jersey, 45, 68
        New Plymouth, 70
        New York, 4, 44, 45, 68, 69, 100, 112
        North Carolina, 45, 56, 69
        Ohio, 82, 101
        Pennsylvania, 44, 45, 70
        Plymouth Colony, 68
        South Carolina, 56
        Tennessee, 94
        Texas, 13
        Vermont, 88
        Virginia, 69, 79, 129
        West Jersey, 44
        West Virginia, 69
    English, 129–130
        Notation, 12
        Æthelred Unrædy, 73
        (ca. 960) 3 Edgar c. 8, 67, 131
        (1225) 9 Henry 3, *Magna Carta*, 80, 130, 131
        (1266) 51 Henry 3, 71, 111, 122–127, 131
        (1301) 31 Edward 1, 71, 73, 74, 79, 82, 89, 90, 111, 122–127, 129, 131
        (1301) 33 Edward 1 stat. 6, 5, 130, 131
        (1325) 18 Edward 2, 28
        (1335) 9 Edward 3 stat. 1, 87, 131
        (1340) 14 Edward 3 stat. 1 c. 12, 130, 131
        (1340) 14 Edward 3 stat. 1 c. 21, 89, 131
        (1341) 15 Edward 3 stat. 3 c. 3, 89, 132
        (1350) 25 Edward 3 stat. 5 c. 1, 132
        (1350) 25 Edward 3 stat. 5 c. 9, 132, 134
        (1350) 25 Edward 3 stat. 5 c. 10, 130, 132
        (1353) 27 Edward 3 stat. 1 c. 8, 48, 132
        (1353) 27 Edward 3 stat. 2 c. 10, 87, 132
        (1354) 28 Edward 3 c. 14, 4
        (1357) 31 Edward 3 stat. 1 c. 2, 132
        (1357) 31 Edward 3 stat. 1 c. 5, 132
        (1360) 34 Edward 3 c. 5–6, 132
        (1363) 37 Edward 3 c. 7, 132
        (1373) 47 Edward 3 c. 1, 132
        (1380) 4 Richard 2 c. 1, 48, 133
        (1381) 5 Richard 2 c. 4, 133
        (1389) 13 Richard 2 stat. 1 c. 9, 89,

Index

130, 133, 134
(1389) 13 Richard 2 stat. 1 c. 10, 133
(1390) 14 Richard 2 c. 11, 133
(1391) 15 Richard 2 c. 4, 133, 141
(1392) 16 Richard 2 c. 1, 133
(1405) 7 Henry 4 c. 10, 133
(1409) 11 Henry 4 c. 6, 133
(1411) 13 Henry 4 c. 4, 133
(1413) 1 Henry 5 c. 10, 133, 134, 141
(1414) 2 Henry 5 stat. 2 c. 4, 39, 133
(1421) 9 Henry 5 stat. 1 c. 11, 133
(1421) 9 Henry 5 stat. 2 c. 6, 72, 133
(1423) 2 Henry 6 c. 14, 44, 47, 55, 129, 134
(1423) 2 Henry 6 c. 15, 102
(1423) 2 Henry 6 c. 16, 39, 134
(1429) 8 Henry 6 c. 5, 134
(1430) 9 Henry 6 c. 8, 89, 129, 134
(1433) 11 Henry 6 c. 8, 130, 134, 141
(1433) 11 Henry 6 c. 9, 134
(1439) 18 Henry 6 c. 16, 134
(1439) 18 Henry 6 c. 17, 47, 134
(1464) 4 Edward 4 c. 1, 134
(1468) 8 Edward 4 c. 1, 134
(1482) 22 Edward 4 c. 2, 48, 55, 134
(1483) 1 Richard 3 c. 8, 134
(1483) 1 Richard 3 c. 13, 47, 129, 134
(1491) 7 Henry 7 c. 3, 135
(1491) 7 Henry 7 c. 7, 48, 135
(1494) 11 Henry 7 c. 4, 80, 89, 135
(1494) 11 Henry 7 c. 23, 48, 55, 135
(1496) 12 Henry 7 c. 5, 25, 45, 48, 64, 79, 111, 122–127, 129, 130, 135
(1503) 19 Henry 7 c. 1, 4
(1514) 6 Henry 8 c. 9, 89, 135
(1529) 21 Henry 8 c. 12, 90, 135
(1531) 23 Henry 8 c. 4, 56, 102, 129, 135
(1532) 24 Henry 8 c. 3, 28, 90, 135
(1536) 28 Henry 8 c. 14 §5, 48, 135
(1570) 13 Elizabeth 1 c. 11 §4, 55, 135
(1581) 23 Elizabeth 1 c. 8 §4, 56, 122, 135
(1589) 31 Elizabeth 1 c. 8, 135
(1593) 35 Elizabeth 1 c. 6 §8, 4, 135
(1601) 43 Elizabeth 1 c. 14, 135
(1640) 16 Charles 1 c. 19 §2, 136
(1640) 16 Charles 1 c. 19 §6, 136
(1660) 12 Charles 2 c. 23 §8, 136
(1660) 12 Charles 2 c. 24 §21, 136
(1662) 14 Charles 2 stat. 2 c. 26 §1, 136
(1664–65) 16 & 17 Charles 2 c. 2 §1, 136
(1670) 22 Charles 2 c. 8 §2, 83, 136
(1670–71) 22 & 23 Charles 2 c. 12, 136
(1688) 1 William & Mary stat. 1 c. 24 §4, 136
(1693) 5 William & Mary c. 7 §9, 136
(1695–96) 7 & 8 William 3 c. 31 §43, 136
(1696–97) 8 & 9 William 3 c. 22 §45, 70, 136
(1697–98) 9 William 3 c. 6 §1, 136
(1698) 10 William 3 c. 10 §12, 136
(1698–99) 11 William 3 c. 15, 136
(1701) 13 & 14 William 3 c. 5 §9, 136
(1702) 1 Anne stat. 1 c. 9 §1, 137
(1702) 1 Anne stat. 1 c. 15 §6, 137
(1702) 1 Anne stat. 1 c. 15 §9, 137
(1702) 1 Anne stat. 2 c. 3 §6, 137
(1706) 6 Anne c. 11 art. 8, 137
(1706) 6 Anne c. 11 art. 17, 137
(1706) 6 Anne c. 27 §22, 46, 47, 64, 98, 137
(1709) 8 Anne c. 10 §1, 137
(1710) 9 Anne c. 6 §5, 137
(1710) 9 Anne c. 6 §8, 137
(1710) 9 Anne c. 17, 137
(1710) 9 Anne c. 20, 137
(1711) 10 Anne c. 43, 137
(1713) 13 Anne c. 20 §13, 137
(1735) 8 George 2 c. 12 §2, 137
(1751) 24 George 2 c. 31 §8, 137
(1775) 15 George 3 c. 27, 137
(1797) 38 George 3 c. 89, 56
(1824) 5 George 4 c. 74, 65
Statutes at Large, see Pickering, D.
Statutes at Large of Pennsylvania, 64
Statutes of Ohio and of the Northwestern Territory, 82
Statutes of the Mississippi Territory, 94
Statutes of the Realm printed by command of George III, 130
Steel, 26

Sterling penny, *see* Tower weight
Sterling silver (alloy), 132, 133
Stone
    Avoirdupois, *see* Avoirdupois weight
    For flax, 137
    For hemp, 135, 137
Stonehenge, 58
Strasburgh, 85
Stratmann, F. H., *Middle-English Dictionary*, 39
Striked (or struck) measure, 50, 59, 67, 129, 132, 133, 136, 141
Strikes, 59
Strong water, 70
Stroud, R., 39, 66
Suakin, 76
Subtle wares, 93
Sudan, 33
Sulu Archipelago, 37
Sumerian measures, *see* Mesopotamian measures
Sumero-Babylonian metrology, 1–2
Survey foot, 12
Survey yard, 12
Svalöv, Sweden, 114–116
Svenska Cereallaboratoriet AB, 127, 128
Swan, *Statutes of the State of Ohio*, 104
Sweden, 128
Syria, 9, 34
Syrian mina, 23, 29, 77
Syrian talent, 23, 35, 75, 78
Syrian/Phoenecian kotyle, 78
System vs. standard, 41, 72

T'sun, 10, 78
Tablespoon, 42, 148
Tael, *see* Asian weights
Talent, *see* Ancient weights
Tallow, 26, 131
Talmud, 63
Tar, 45, 56, 64, 69
Tare, 93
Tarkhan, 21
Tauris, 31
Teaspoon, 41, 148
Tehran, 31
Tennessee, 94
Tertian, *see* Cask measures

Test Weight and Your Next Wheat Crop, 127
Test weight of grain, 111–127
Test weights of generic wheat and barley, 112–114
Test weights of specific varieties of grain, 114–118
Tetradrachma, 84, 148
Texas, 13
Thailand, *see* Siam, 38
Thangsat (Siam), 55
The old measure, 47
"The whete eare. Too graynes maketh ye xvi pte of a penny.", 72
Thom and Thom, *Megalithic Remains in Britain and Brittany*, 65
Thom, A.
    "The Geometry of Cup-and-Ring Marks", 65
    Megalithic Sites in Britain, 65
Thompson v. District of Columbia, 21 AD. C. 395 (1904), 129
Thread, 26
Thrimsa or thrymsa, *see* Trimes
Thumb, *see* Inch, 148
Thuoc moc, 54
Thureau-Dangin, F., 54, 65
    "L'u, le Qa et la Mine,", 65
    "La Mesure du qa", 66
    "Numération et Métrologie Sumériennes", 2, 65
    "Sketch of a History of the Sexagesimal System", 2, 12, 38
Tierce, *see* Cask measures
Tin, 26, 62, 123
Tod, *see* Avoirdupois weight; Wool weights
Toiletries, 43
Tomb of Hesy, 59–62, 78, 79
Toulmin, H., *Digest of the Laws of the State of Alabama*, 65
Tower weight, 133
    Merchant's pound, 73, 75, 77, 90, 102, 111, 122–127, 131
    Ounce, 30, 71–76, 78, 131
        Definition, 71
    Pound, 71–75, 77, 131
        Definition, 71
    Pound of Cologne, 29–30, 73, 77
    Saxon graves, 75

Index

Shilling, 73
Sterling penny, 71–76, 131, 145
    Definition, 71
Stone of London, 71–72
    Definition, 71
Tower wheat corn, 71–74, 131
Tractatus de Ponderibus et Mensuris (Treatise on Weights and Measures, 31 Edward 1, 1301), 71, 73, 74, 79, 82, 89, 90, 111, 122–127, 129, 131
Trash barrels, 56
Treatise on Weights and Measures, *see* Tractatus de Ponderibus et Mensuris
Treaty of 1560, 29
"Trial of weights and measures under Henry VII", 81
Trimes or trims, 74
Tripoli, 31
Troy weight, 15–38, 62, 133
    Ancient history, 20–25
    Destruction in Britain, 16–17
    Egyptian origins, 21–23
    Etymology of *Troy*, 24
    Grain, 16, 18, 123, 124, 142
    Haverdepois ounce, 28
    Haverdepois pound, 26–29, 89–91, 135
    In Asia, 36–38
    In India, 35–36
    Inherited from Arabs, 30
    Ounce, 16, 18, 45, 75, 107, 123, 124
        In Egypt, 21
    Pennyweight, 16, 18, 25, 45, 123, 124, 141, 146
    Pound, 15–38, 45, 75, 77, 123, 124, 134
        Argive, 23
        Definition, 25, 111, 122–127, 135
        In Egypt, 21
        In Japan, 38
        In Laos, 38
        Parliamentary Standard of 1758, 16–18
        Standard of the Philadelphia Mint, 18
    Pound of Cologne, 29–30, 73, 77, 91
    Shilling, 19, 20, 26
    Sterling, 25, 45
    Survival in U.S., 17–18
    Troy wheat corn, 25, 45
Troyes, 24, 85

Tun, *see* Cask measures
Tun for coal, 137
Tunis, 31
Turkey, 9, 21
Turpentine, 56, 64

Ukiah (okia, vachia, etc.), *see* Arabic weights
Ulna, 141
Uluburun shipwreck, 22
Uncia, 20
Unit fractions, 53
Unit pricing, 108–109
United Nations, 30
United States v. Weber, 3 Ct. Cust. A19 (1912), 129
University of Chicago, 57
University of Nebraska, 119
Ur-unit of 48 grains, 21
U.S. Animal and Plant Health Inspection Service, 119
U.S. Congress, Joint Resolution of 1836, 101
U.S. Constitution, 129
U.S. Customs Houses, 104
U.S. Department of Commerce, xi
    Units and Systems of Weights and Measures (LC 1035), 12, 69
U.S. half dollar coin, 38
U.S. No. 1 wheat, 46, 73, 112, 119, 122
U.S. peck, dry gallon, dry quart, dry pint, *see* Winchester measure
U.S. wheat and barley, 1902–1931, 113
USDA Federal Grain Inspection Service, 128
USDA *Grain Inspection Manual*, 127
USDA *Weather, Crops, and Markets*, 127

Vaat (Estonia), 49
Vakia, 59
Vara, 9
Vase of Entemena, 57, 62
Veal, 28, 90, 135
Vegetables, 67
Velte (France), 48
Vermont, 88
Vienna, 75, 84
Viertel (Saxony), 79
Vietnam, 38
Viking/Irish weights, 25, 75, 76
Vinegar, 48, 136

Virgate, 5, 149
Virginia, 69, 79, 129
Visus Frankiplegii (View of Frankpledge), 28

Washington, 125
Washington, G., 107
Wax, 26, 123
Weight of grain, 111–127
Weight of water, 45, 51, 100–101
Weight system diagrams
    Ancient Attic and medieval tower weight systems, 75
    Arabic troy survivals, 35
    Dominant bronze-age system, 23
    Early Greek weight systems, 23
    Synthetic overview of primary ancient and medieval standards of weight, 78
    Troy weight in India, 36
    Troy weight in Saxon and medieval England, 29
Weight vs. capacity, *see* Connection between volume/capacity and weight
Weights and densities, 77–78
"Weights and Measures of the City of Winchester", 82
Wessex, 57
West Jersey, 44
West Virginia, 69
Wey of cheese, 134
Wheat, 20, 45, 73, 77, 87, 96, 111–127
    "Best quality", 119–122
    Grading, 112–127
    Height, 116–117
    Spring, 113, 116
    Test weight, 46, 111–127
        Cornell, 116–118
        Export statistics, 119, 122
        Generic, 112–114
        International Winter Wheat Performance Nursery, 119–122
        London, 1834, 119, 122
        Paris, 18th century, 119, 122
        Specific varieties, 114–118
        Svalöv, Sweden, 114–116
        U.S. No. 1, 46, 73, 112, 119, 122
        Versus yield, 117
    Winter, 113, 116–117, 119–122
    Yield, 116–117, 119–122
Wheat corn, 46
    "From the middle of the ear", 119
    Tower weight, 71–74, 131
    Troy weight, 25, 45
Wheat density vs. wine density, 77–78
Whole-number ratios between systems, 53
Whole-number volume ratios
    Cubic Egyptian cubit and cubic Sumerian foot, 54
    Cubic Northern foot and cubic Greek foot, 54
Width vs. heap, 67
William of Malmesbury, 95
Winchester, 8, 28, 68, 131
Winchester measure, 67–82
    Barrel, 68–69, 79
    Barrel for fruits and vegetables, 69
    Bushel, 45, 67, 68, 70–72, 97, 100, 111, 122–127, 129, 131, 133, 136, 140
        Apples, 129
        Definition, 68, 70, 71, 136, 137
        Dimensions, 70
        Exchequer standard of Elizabeth I, 71
        Exchequer standard of Henry VII, 71, 79
    Chaldron, 68, 140
    Cornook, 123
    Demibushel, 68
    Faat (heaped quarter), 133, 134, 141
    For liquids, 69–70
    Gallon, 41, 44, 55, 64, 67, 68, 70–72, 101, 123, 124, 131
        Exchequer standard of Elizabeth I, 71
        Exchequer standard of Henry VII, 71
    Half bushel
        Dimensions, 83
    Half peck, 68
        Dimensions, 83
    Heaped, *see* Heaped measures
    Last, 123
    Peck, 67, 68, 123, 124, 132, 145
        Dimensions, 83
    Pint, 67–70, 123
    Pottle, 132
    Quart, 67–70, 132
    Quarter, 68, 71–72, 123, 131, 133, 135, 146

Index

    Saxon origins, 78–79
    Stone, 135
    Strike, 123
    "Water measure" (five pecks), 135, 136
        Redefined as the bushel, 137
Winchester quart and pint (so called), 70
Wine, 46–48, 59, 66, 70, 77, 87, 96, 105, 106, 111–127, 131–135
    Specific gravity, 85
    Weight, 111–127
Wine gallon, *see* Gallon
Wine glass, 42, 149
Wisconsin, 64
Woket, *see* Arabic weights
Wood, 26
Wool, 73, 87
Wool weights, 15, 26, 88–91, 131–133, 135, 147, 148
Wrenches, 106
Wright, J., *English Dialect Dictionary*, 39
Württemberg, 8, 79
Wycliffe Bible, 12
Wylie, C. C., "On King Solomon's Molten Sea", 66

Yang, *see* Tael
Yard, 68, 149
    Customary, 5–12, 95–103, 106, 131, 132
    Exchequer standard of Elizabeth I, 97
    Exchequer standard of Henry VII, 97
    Megalith, 58
    Merchant Taylors' Company, 95, 102
    Northern, 8, 11, 106
    Saxon, 96
    Standard of 1760, 16
    Survey, 12
    Winchester City, 95
"Yard and hand," "yard and inch," "yard and nail," "yard and thumb", 134
York Memorandum Book, 85

Zambaur, E. V., "Dirham", 39
Zanzibar, 31
Zuidhof, A., 66
Zupko, R. E., *British Weights & Measures*, x, 82, 104